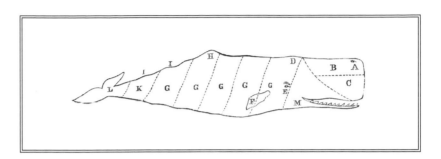

Trying Leviathan

The Nineteenth-Century

New York Court Case

That Put the Whale on Trial

and Challenged the Order of Nature

D. GRAHAM BURNETT

PRINCETON UNIVERSITY PRESS

Princeton & Oxford

In the United Kingdom: Princeton University Press, 3 Market Place,
Woodstock, Oxfordshire OX20 ISY

Library of Congress Cataloging-in-Publication Data
Burnett, D. Graham.
Trying Leviathan : the nineteenth-century New York court case that put
the whale on trial and challenged the order of nature / D. Graham Burnett.
p.   cm.
Includes bibliographical references and index.
ISBN 978-0-691-12950-1 (hardcover : alk. paper)
1. Maurice, James, fl. 1818—Trials, litigation, etc. 2. Judd, Samuel, fl. 1818—
Trials, litigation, etc. 3. Trials (Tax evasion)—New York (State)—New York—History—
19th century. 4. Whale oil—Taxation—Law and legislation—New York (State)—History—
19th century. 5. Whales—Classification—History—19th century. 6. Whaling—United
States—History—19th century. 7. Zoology—Social aspects—United States—History—
19th century. I. Title.
KF228.M38B87 2008
344.747'0957—dc22
2007008378

British Library Cataloging-in-Publication Data is available

This book has been composed in Adobe Caslon

Printed on acid-free paper. ∞

press.princeton.edu

Printed in the United States of America

1   3   5   7   9   10   8   6   4   2

For Francesca Socorro

When Leviathan is the text, the case is altered.

HERMAN MELVILLE, *Moby-Dick*

Contents

COLOR PLATES

(following page 110)

FIGURES

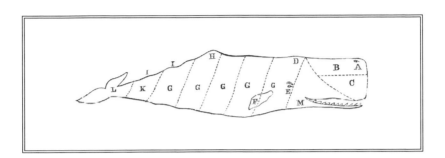

THE PEACE OFFERING THAT STANK

On the 9th of November, 1819 the distinguished New York physician-naturalist, Samuel Latham Mitchill, received at his Barclay Street chambers a malodorous packet containing an over-ripe orange file fish (*Aluterus schoepfii*, Walbaum, 1792) that had run afoul of a Long Island boating party several days earlier and had succumbed to "the stroke of an oar" on the beach near Bowery Cove. This rank offering arrived in the company of an explanatory letter from the Irish Jacobin lawyer William Sampson, the doctor's fellow member of the Literary and Philosophical Society of New-York, and his sometime adversary in the courts of the city.

"The fish begins to have an ancient and more than fish like smell," wrote Sampson by way of greeting, and he continued, "I send it to you, that you may either pass judgement upon it or record its characters before it undergoes more alteration." Could the specimen be a hitherto unknown species, something strange and new? To assist Mitchill on this question, Sampson had gone so far as to enclose a "colored drawing" prepared by his daughter to capture the creature's appearance in its fresher state.[1]

Dr. Mitchill knew better, and he made a terse note overleaf, identifying the genus and species. Not only was it not new, Mitchill himself had already published on it, and in his annotation on the letter he cited his 1815 monograph on "The Fishes of New-York," writing across the covering flap, " . . . already described in the Lit. & Ph. Transactions of NY."

It was, in a way, a familiar ritual. Mitchill's ichthyological knowledge was, by 1819, legendary in New York and beyond, so much so that he

---

[1] Letter from William Sampson, 9 November 1819, Gratz Collection, Historical Society of Pennsylvania. The addressee of this letter has not previously been identified. It is worth noting that Sampson's salutation is an oblique reference to *The Tempest*, act 2, scene 2, where Trinculo, examining Caliban, declares, "A fish; he smells like a fish; a very ancient and fish-like smell; a kind of not-of-the-newest Poor-John [i.e., salted hake]. A strange fish!" The allusion is a specimen of the relentlessly referential eloquence of Sampson and Mitchill both. A general note on citations in this book: spelling has not been modernized in quotations from original sources, but punctuation has in some cases (addition of apostrophes in contractions, etc.) been brought into line with current conventions.

could be satirized in doggerel in the New York papers as the "Phlo'bom-bos of our Icthyology [*sic*]," a title cobbled out of Mitchill's inelegant (if erudite) neologism for Robert Fulton's wondrous new invention, the steamboat, which the doctor proposed might properly be known as the "phlogobombos."[2] Moreover, Mitchill's boosterism for natural-historical investigation in the young Republic had garnered him an expansive circle of correspondents who frequently sent him unusual animals, plants, and rocks as accessions to his collections, or for classificatory consultations.[3] Mitchill—who would go on to be memorialized as the "Nestor of American Science"—had lectured on zoology at Columbia College and at the College of Physicians and Surgeons from their inceptions, and he was famous for offering his students a Cuvier-esque declaration of his taxonomic mastery of sea creatures, announcing boldly to them, "Show me a fin, and I will point out the fish."[4] Nor was this simply the encyclopedism of a cabinet natural historian, one who hid among his dusty books and old specimens: Mitchill was well known for taking his ichthyology to the street. A familiar of the New York fish market, he brought his students to the docks to practice their skills of identification on the catch of the day, and he boasted that more than half of the new fish species he had discovered could be found in that piscatorial emporium.[5]

[2]Joseph Rodman Drake, *Poems, by Croaker, Croaker and Co. and Croaker, Jun.* (New York: For the Reader, 1819), pp. 26–27.

[3]Courtney Robert Hall, *A Scientist in the Early Republic: Samuel Latham Mitchill, 1764–1831* (New York: Columbia University Press, 1934), p. 83; Mitchill to DeWitt Clinton, 20 February 1826, DeWitt Clinton Papers, Columbia University Rare Books and Manuscript Library. See also the papers of Jeremy Robinson (in the Peter Force Collection, Manuscript Division, Library of Congress), a commercial agent in South America in the early nineteenth century, who corresponded with Mitchill.

[4]The French naturalist par excellence, Georges Cuvier, was famous at the time for his boasted ability to identify whole animals from a mere tooth or stray bone. The anecdote in connection with Mitchill can be found in John W. Francis, *Old New York, or, Reminiscences of the Past Sixty Years* (New York: C. Roe, 1858), p. 94. Interestingly, there is evidence that he was taken up on this challenge. The Columbiana Collection in the Rare Books and Manuscript Library at Columbia University (s.v. Mitchill) contains a letter from Mitchill to Edward Neufville (5 July 1822) identifying a species of tropical herring from several scales forwarded by a curious correspondent. Mitchill received a good deal of ribbing in the press concerning this vaunted skill, for instance: "'*Ex pede Herculem*,' said the ancient artist, 'I can tell Hercules from his toe'—'*Ex dente animal*,' says Cuvier, 'give me the bone, and I will describe the animal'—'I go further,' says the immortal Mitchill, 'show me a single scale, and I will let you know the fish that owned it.'" See *American*, 20 November 1820, p. 2.

[5]Charlotte M. Porter, *The Eagle's Nest: Natural History and American Ideas, 1812–1842* (Tuscaloosa: University of Alabama Press, 1986), p. 34. See also Samuel Latham Mitchill, *Picture of New-York, or, The Traveller's Guide through the Commercial Metropolis of the United States* (New York: I. Riley, 1807). For Mitchill giving instruction in the Market, see Reuben Haines to

But this particular specimen of New York waters represented more than just another missive from some friendly fisherman. This fish, rightly understood, was nothing less than a subtle peace offering from the legal establishment of the city. For that boating party on Bowery Cove had included not only the highly visible attorney, belletrist, and serial pamphleteer Sampson, but also Richard Riker, the City Recorder, who sat as the presiding official in most of the trials held in the Mayor's Court of New York City. Together these two men had elected to deliver their odd catch up to Mitchill "as a tribute to science and a token of continuing friendship." Swelling to oratorical excess (as he was wont to do), Sampson, Mitchill's "humble servant," sang the praises of New York's philosopher-king:

> What an empire is that of a man of learning. The King of England by his prerogative gets royal fish and very few indeed of them. You get from all quarters the willing tribute of sea and land, and I sincerely hope this that we send may be an acquisition to you and an addition of the stock of knowledge you have treasured up for mankind, and with which you have so much enriched science and the arts.[6]

It was an elegant rhetorical flourish, this invocation of the "royal fish" of the English Crown, and a delicately oblique gesture at the subtext of Sampson's letter. The "royal fish" was, as any Edinburgh-trained ichthyologist (or Lincoln Inn lawyer) knew, the whale, which had been the special possession of the throne since the fourteenth century.[7] For the English, a king, and for that king, tribute in the form of each beached whale; for the citizens of the land Mitchill had dubbed "Fredonia," no royal prerogatives, only the preeminence of Jefferson-style political sages (like Mitchill, himself a former senator and a state representative, with almost a decade of service in Washington), men whose leadership would forever be founded on an intimacy with the natural productions of

---

James Pemberton Parke, 25 February 1812, folder 128, series ii, Ruben Haines Papers, American Philosophical Society. See also Simon Baatz, "Knowledge, Culture, and Science in the Metropolis: The New York Academy of Sciences, 1817–1970," *Annals of the New York Academy of Sciences* 584 (1990), pp. 1–269, at p. 17.

[6] Letter from William Sampson, 9 November 1819, Gratz Collection, Historical Society of Pennsylvania.

[7] Whales and sturgeon shared this title. The term "Fishes Royal" first appears in a statute of 1324. See Adriaen Coenen, *The Whale Book: Whales and Other Marine Animals as Described by Adriaen Coenen in 1585*, edited by Floike Egmond and Peter Mason (London: Reaktion, 2003), p. 6.

American shores. For Mitchill, then, philosophical tribute in the form of a wizened file fish from the Fredonian strand.

Under ordinary circumstances the courtesy might have seemed strained (if not odd), the conceit labored. But Mitchill would have had no difficulty understanding the allusion: Sampson was offering, in the stinking packet, in the elaborate invocation of the *Praerogativa Regis,* a small token of deference to the man who had recently become the butt of considerable public ridicule where whales and fish were concerned, ridicule occasioned by a very public zoological showdown with Sampson himself.

For on the 30th of December of 1818 Mitchill had appeared in the packed chambers of the Mayor's Court in City Hall as the star witness in the case of *James Maurice v. Samuel Judd,* a dispute arising under a New York State statute that obliged purveyors of "fish oils" to ensure that their casks had been gauged, inspected, and certified. *Maurice v. Judd* had taken shape earlier that year, in the sweltering summer, when the well-known candle maker and oil merchant, Samuel Judd, had refused to pay the inspector's fee on three casks of spermaceti oil—protesting that the oil was not "fish oil" but "whale oil" and that whales were not, in fact, fish. The inspector, James Maurice, gave a derisive snort (whales not fish? ha!) and issued Judd a summons. The stage was thus set for a legal action that would ballot twelve sworn jurors to determine whether, in the state of New York, a whale was a fish. Sampson represented the plaintiff, Maurice, who was suing to collect the statutory fines. Mitchill, ichthyologist extraordinaire, appeared as the heart of the defense, carrying the standard of modern taxonomy. The recorder, Richard Riker, presided as judge. This two-day trial, which became a pageant of natural historical erudition and a sensational *agon* for settling natural and social order, would represent a remarkable instance of science at the bar in the nineteenth century, and it would live in popular memory and in the works of classifying naturalists for decades, not least because of William Sampson's transcript of the trial, *Is a Whale a Fish? An Accurate Report of the Case of James Maurice against Samuel Judd.*[8] This pamphlet appeared in bookshops in the summer of 1819.

The autumn of 1819 was thus a very suitable moment for the author of *Is a Whale a Fish?* (which would open Mitchill to new ribbing) to make gracious gestures toward one of the most powerful and politically connected men of learning in the city of New York.

---

[8]William Sampson, *Is a Whale a Fish? An Accurate Report of the Case of James Maurice against Samuel Judd* (New York: Van Winkle, 1819), henceforth abbreviated *IWF.*

MAURICE V. JUDD AND THE HISTORY OF SCIENCE

This book takes up the unusual case of *Maurice v. Judd*, reconstructing the trial itself from available published and manuscript sources, rehearsing (wherever possible) the biographies of the persons involved, tracing textual allusions and references, situating the legal, political, and scientific arguments made in the proceedings, and detailing the public response to, and enduring legacy of, this event. Broadly, I aim to recover the trial and reveal its larger historical significance.

Why bother? This case merits the reader's attention for three reasons. First (and most narrowly), *Maurice v. Judd* represents a telling episode in the history of science in the early Republic, and in New York in particular. Where New York itself is concerned, the trial sheds light on the status of "philosophy" (and "philosophers") in general, and natural history specifically, during critical years in the emergence of the city's learned institutions and intellectual culture.[9] Broadening the focus to the Republic as a whole, the case invites us to revise a dominant theme in the literature treating the history of the natural sciences in the United States in the first half of the nineteenth century, a theme that relentlessly emphasizes the way that natural history served as a tool (and proxy) for an emerging "American identity" rooted in a kind of nature-nationalism.[10] Qualifications to this thesis are overdue.

[9]The best general treatment of the history of science in the United States in this period remains John C. Greene, *American Science in the Age of Jefferson* (Ames: Iowa State University Press, 1984). Particularly relevant is chapter 4, "Science along the Hudson." Philip Pauly has focused on the natural sciences in the first chapter of his *Biologists and the Promise of American Life* (Princeton: Princeton University Press, 2000). Andrew John Lewis's dissertation ("The Curious and the Learned: Natural History in the Early Republic," Yale University, 2001) offers several detailed and helpful case studies. On the general issue of intellectual culture in New York City to 1830, see part I of Thomas Bender's *New York Intellect: A History of Intellectual Life in New York City from 1750 to the Beginnings of Our Own Time* (New York: Knopf, 1987). On science and learned societies in New York in these years: Baatz, "Knowledge, Culture, and Science"; Jonathan Harris, "New York's First Scientific Body: The Literary and Philosophical Society, 1814–1834," *Annals of the New York Academy of Sciences* 196 (1972), pp. 329–337; Brooke Hindle, "The Underside of the Learned Society in New York, 1754–1854," in *The Pursuit of Knowledge in the Early American Republic: American Scientific and Learned Societies from Colonial Times to the Civil War*, edited by Alexandra Oleson and Sanborn C. Brown (Baltimore: Johns Hopkins University Press, 1976); Kenneth R. Nodyne, "The Founding of the Lyceum of Natural History," *Annals of the New York Academy of Sciences* 172 (1970), pp. 141–149; and idem, "The Role of De Witt Clinton and the Municipal Government in the Development of Cultural Organizations in New York City, 1803–1817," Ph.D. dissertation, New York University, 1969.

[10]I am thinking here of works like David Scofield Wilson's *In the Presence of Nature* (Amherst: University of Massachusetts Press, 1978) and, more recently, Paul Semonin's "'Nature's Nation': Natural History as Nationalism in the New Republic," *Northwest Review* 30 (1992), pp. 6–41, and Christopher Looby, "The Constitution of Nature: Taxonomy as Politics

Second, this case offers a unique point of departure for a more general consideration of cetaceans (whales, dolphins, and porpoises) as "problems of knowledge" in the early nineteenth century. In these years whaling in the open ocean was rapidly becoming the young Republic's strongest claim to global preeminence and indefatigable enterprise. As early as 1775 Edmund Burke had pontificated in Parliament that the Yankee whalemen were sweeping the globe, and humiliating British seafarers with their daring.[11] By the 1840s some 600 American whaling vessels were plying the Pacific, vanguards of U.S. geopolitical ambitions, and a major source of national wealth.[12] In short, whales mattered to the early United States. And at stake in the trial of *Maurice v. Judd* was the essential nature of these unusual (and economically vital) animals—their form, habits, and place in the natural order. For this reason the testimony of different witnesses—European-educated men of science like Mitchill, New England whalemen, merchants and agents in the whaling industry, artisans and craftsmen accustomed to work with whale products—provides unique insights into who knew what about these creatures, and how they authorized their claims. In this book, then, the trial will afford the occasion for several brief departures, detours, loops out and away from the courtroom, in which we pause to consider, for instance, what whalers knew about the anatomy, physiology, and natural history of their quarry during this period. Other similar issues will receive attention: Where could New Yorkers (like the members of the jury) have seen whales or whale parts in 1818? What was the status of the whale in parlors, primers, and schools in the period?[13] Such digressions and amplifi-

in Jefferson, Peale, and Bartram," *Early American Literature* 22, no. 3 (1987), pp. 252–273. It is perhaps unfair to cherry-pick, but this quote is illustrative of the general claim I suggest we consider more closely: "The static displays of Peale's collection (and, within it, the mammoth, at its summit and center) reiterate the nation-building, class-constituting technique of collective incorporation through production and its display, or the process of assembly itself preserved and restored as the grounds of *E pluribus unum*." Laura Rigal, *The American Manufactory: Art, Labor, and the World of Things in the Early Republic* (Princeton: Princeton University Press, 1998), p. 94.

[11] The much-quoted speech reads in part: "whilst we are looking for them beneath the Arctic Circle, we hear that they have pierced into the opposite region of polar cold, that they are at the antipodes . . . whilst some of them draw the line and strike the harpoon on the coast of Africa, others run the longitude, and pursue their gigantic game along the coast of Brazil. No sea but what is vexed by their fisheries." Quoted in Jeremiah N. Reynolds, *Address, on the Subject of a Surveying and Exploring Expedition* (New York: Harper and Brothers, 1836), p. 7.

[12] For this story, see D. Graham Burnett, "Hydrographic Discipline," in *The Imperial Map*, edited by James Akerman, forthcoming from University of Chicago Press.

[13] The phrase is an allusion to Sally Gregory Kohlstedt, "Parlors, Primers, and Public Schooling: Education for Science in Nineteenth-Century America," *Isis* 81, no. 3 (1990), pp. 424–445.

cations will help contextualize the trial itself, even as they offer opportunities for a walk though the world of learning in the early Republic.

Third, and perhaps most importantly, *Maurice v. Judd* will serve as a window onto the contested territory of zoological classification in the late eighteenth and early nineteenth centuries. Here whales are only part of the story, albeit a more important part than has generally been recognized. According to a dominant narrative in the history of science, the second half of the eighteenth century represents the golden age of the classifying imagination, the period during which, under the presiding spirit of Linnaeus, nomenclature and systematic taxonomy reduced the oozy, Protean organism of the Renaissance world to a proper natural order, an order in which things had names and places in schematic hierarchies that were themselves (to a greater or lesser degree) reflections of the nature of things.[14] By the early nineteenth century, as this account goes, the labors of the enlightened classifiers had settled the lineaments of this natural tableau, and in so doing created the conditions of possibility for a series of iconoclastic and revisionist theories of the living and non-living world—theories less wedded to fixity and rigid types, and more interested in accounting for time, change, and the genealogy of a world increasingly seen as deeply, rather than superficially, contingent (cue the Darwinian revolution). In an imaginative series of essays published as *The Platypus and the Mermaid,* Harriet Ritvo has worked to destabilize this familiar account of the "heroic age of scientific classification."[15] It is her claim that historians of science have largely overlooked "quite a different zoological enterprise—one in which consensus was rare, in which authority was uncertain and fragmented, and in which the very principles behind the construction of taxonomic systems and the assignment of individual species to their niches were vaguely defined and of obscure or questionable provenance."[16] By focusing on problematic cases and anomalous organisms, and by situating the debates of learned classifiers in the broader context of lay expertise among turn-of-the-century English animal breeders, hunters, farmers, and fanciers, Ritvo

[14]The now classic (if contested) statement is that of Michel Foucault (*The Order of Things: An Archaeology of the Human Sciences* [New York: Pantheon Books, 1971]), who drew heavily on the much earlier work of Henri Daudin (*Cuvier et Lamarck: Les Classes Zoologiques et l'Idée de Série Animale 1790–1830* [Paris: F. Alcan, 1926]). Historians of science have never rested easy with Foucault's analysis, even if it did have the "merit of awakening historians from their dogmatic slumbers." See the rich work by James L. Larson, *Interpreting Nature: The Science of Living Form from Linnaeus to Kant* (Baltimore: Johns Hopkins University Press, 1994), p. 3.

[15]Harriet Ritvo, *The Platypus and the Mermaid, and Other Figments of the Classifying Imagination* (Cambridge, MA: Harvard University Press, 1997), p. 14.

[16]Ibid., p. 19.

is able to suggest that, as she puts it, "a great deal remained up for grabs" in the period:

> To participants in a project conventionally hailed as demonstrating the intellectual conquest of nature, the internal history of zoological classification might seem as much a constantly shifting kaleidoscope of competing systems and principles, as a steady evolution and collaboration of a dominant paradigm.[17]

Ritvo's discussion of broader public resistance to Linnaean classification certainly enriches our sense of the period, revealing that, to many literate Britons, the classificatory sciences looked less like philosophy rampant than philosophy rudderless. But she has still bigger fish to fry: it is her more ambitious assertion that the "membrane" between specialist and lay communities was, as she puts it, "highly permeable in both directions."[18] To support this argument she gathers evidence that certain significant categories of analysis—for instance "savage" and "domestic"—amounted to "taxonomic differentia smuggled into systematic zoology from a lay world of utilitarian and anthropocentric discrimination."[19] To the degree that she succeeds here, Ritvo's "bottom up" history of systematics does more than provide a nuanced cultural history of the classificatory sciences in pre-*Origin* Britain; it has the potential to shed valuable light on issues central to the content of the sciences of life in the nineteenth century—saliently, Darwin's ideas about hybridity and artificial selection.

Maurice v. Judd offers a remarkable opportunity to investigate, for the United States, problems like those taken up by Ritvo in Britain during the same period: the case presents a gloriously feisty public forum where competing parties deployed a wide range of skills, texts, and authorities in efforts to undermine (and sometimes to undergird) the edifice of contemporary taxonomy and classification. Nor are these different positions merely static: by following the citations marshaled by the diverse parties to the action I will show how knowledge of natural order and natural types "migrated" across different communities of expertise, and across geographical regions, thereby revealing how the "new philosophy" of the metropolitan (and largely French) classifying science made its way to American readers, and how such ambitious "systems" fared in confrontation with folk taxonomies, vernacular natural history, and biblical

[17] Ibid., p. 38.
[18] Ibid., p. 45.
[19] Ibid., p. 41.

representations of creation. These trial transcripts thus dramatize just how unstable the science of natural order was in 1818, at least as viewed from lower Manhattan by readers who had access to a preponderance of the leading publications in Anglo-European natural history; indeed, it was by setting these texts against each other that opponents of the "new philosophy" could represent the science of classification as a house woefully divided, and by no means the architecture of the natural world. Having revealed the contingency of such "systems," the skeptics were positioned to defend the legitimacy of the taxonomic discriminations implicit in ordinary language and in the social and political categories precipitated out of labor, law, and the market. The adversarial setting of the Mayor's Court dramatized these conflicts, and for historians of science and scholars of law generally interested in the relationship between legal systems and the production of knowledge, *Maurice v. Judd* is a mini-bonanza.[20] For anthropologists, historians, and scientists concerned to explore how human beings have "thought with animals" at different times and places, there is much here on which to graze.[21]

---

[20] A good point of departure here is Sheila Jasanoff, *Science at the Bar: Law, Science, and Technology in America* (Cambridge, MA: Harvard University Press, 1995). While Jasanoff is primarily concerned with the twentieth century, she shows convincingly how courtroom standards and practices pose serious challenges to the claims of scientific experts (adversarial trials can become, she suggests, "orgies of deconstruction" [p. 53], a phrase that captures something of what takes shape in *Maurice v. Judd*). Other relevant recent studies of law, medicine, and science include: Roger Smith and Brian Wynne, eds., *Expert Evidence: Interpreting Science in the Law* (London: Routledge, 1989); James C. Mohr, *Doctors and the Law: Medical Jurisprudence in Nineteenth-Century America* (Baltimore: Johns Hopkins University Press, 1993); David Delaney, *Law and Nature* (Cambridge: Cambridge University Press, 2003); and Tal Golan, *Laws of Men and Laws of Nature: The History of Scientific Expert Testimony in England and America* (Cambridge, MA: Harvard University Press, 2004). For additional discussion of this area (particularly the range of new studies on the "technologies of truth"), see the articles I assembled and introduced in the recent "Focus" section of *Isis*: Burnett et al., "Science and the Law," *Isis* 98, no. 2 (2007), pp. 310–350.

[21] An irreverent colleague refers to this field as the "new crittercism." For a useful introduction, see Lorraine Daston and Gregg Mitman, eds., *Thinking with Animals: New Perspectives on Anthropomorphism* (New York: Columbia University Press, 2004). Examples from this large literature would include: James C. Turner, *Reckoning with the Beast: Animals, Pain, and Humanity in the Victorian Mind* (Baltimore: Johns Hopkins University Press, 1980); Harriet Ritvo, *The Animal Estate: The English and Other Creatures in Victorian England* (Cambridge, MA: Harvard University Press, 1987); Donna Haraway, *Primate Visions* (London: Routledge, 1989); Roy G. Willis, ed., *Signifying Animals: Human Meaning in the Natural World* (London: Unwin Hyman, 1990); Angela N. H. Creager and William Chester Jordan, eds., *The Animal/Human Boundary: Historical Perspectives* (Rochester, NY: University of Rochester Press, 2002); and Mary Heninger-Voss, ed., *Animals in Human Histories: The Mirror of Nature and Culture* (Rochester, NY: University of Rochester Press, 2002). Historians of science have mined this vein with some success, writing studies of particular taxons and their place in the development of scientific ideas and practices. See, for instance: Robert E. Kohler, *Lords of the Fly: Drosophila Genetics*

Though whales were at issue—lumbering creatures in distant seas—their place in the world could not, in the end, be separated from charged ideas about human beings and the peculiarly solipsistic preoccupations of terrestrial featherless bipeds: as we will see, race, gender, and class were at stake, explicitly, when the citizens of New York had to legislate the order of nature. It is perhaps cliché to assert that all taxonomy is politics, or to insist that epistemological problems are always also problems of social order; *Maurice v. Judd* provides a striking occasion to test the viability (as well as the limits) of such sweeping claims.[22]

But were epistemological problems authentically at stake in the effort to construe an obscure New York State statute entitled, humbly, "an act authorizing the appointment of guagers [*sic*] and inspectors of fish oils"?[23] I will argue that they were. The "new philosophy" that exercised the adversaries in this trial represented a powerful (and novel) technique for ascertaining the relations among living creatures, a technique that was, in the second decade of the nineteenth century, codifying the gains of a rapid ascendancy, and institutionalizing its practices and its claims at the most visible center for natural history in the world, the Muséum National d'Histoire Naturelle in Paris. It was there, under the imperious hand of Georges Cuvier, that "comparative anatomy" had become the indispensable guide to natural order. Since this is, loosely, the "new philosophy" at issue in *Maurice v. Judd*, a brief review of the significance of this development in natural history is in order.

To set these significant changes in high relief, we might begin by recalling that so towering a figure in the world of natural history as Georges-Louis Leclerc, Comte de Buffon, could write in the mid-eighteenth century that anatomy was "a foreign object to natural history . . . or at least

---

and the Experimental Life (Chicago: University of Chicago Press, 1994); Daniel P. Todes, "Pavlov's Physiology Factory," *Isis* 88, no. 2 (1997), pp. 205–246; and Karen Rader, *Making Mice: Standardizing Animals for American Biomedical Research, 1900–1955* (Princeton: Princeton University Press, 2004).

[22] The classic statement of the sociological basis of classification remains: Emile Durkheim and Marcel Mauss, "De Quelques Formes Primitives de Classification," *Année Sociologique* 6 (1903), pp. 1–71. For an invaluable discussion of the critical reception of their argument (together with a spirited defense of its continued centrality to the history and philosophy of science), see David Bloor, "Durkheim and Mauss Revisited: Classification and the Sociology of Knowledge," *Studies in History and Philosophy of Science* 13, no. 4 (1982), pp. 267–297.

[23] See New York (State) Legislature, *Journal of the Assembly of the State of New-York: At Their 42nd Session* (Albany: J. Buel, 1819), p. 263.

not its principal object."[24] And while it is possible to point to a long tradition of animal dissection and anatomical study, such work was not understood, before Cuvier, to be *the* privileged basis for the sciences of classification.[25] On the contrary, as Thomas Henry Huxley quipped at the end of the nineteenth century (recalling the bad old days before comparative anatomy), "the pure systematic zoologist was unaware that the stuffed skins he named and arranged ever had contained anything but straw."[26] In other words, back then innards weren't part of the classifying game.

Debates among zoological practitioners of the classifying sciences in the eighteenth century never yielded perfect consensus on the proper way to access the "natural order" of the animal world (nor indeed were all of them persuaded such a thing was accessible to mere mortals), but the prestige, scale, and greater sophistication of botanical investigation in the period—and, above all, the success of Linnaeus's sexual system for plant taxonomy—tended to draw attention to discrete and visible *external* characteristics of creatures, preferably characteristics that differed neatly by number, kind, or combinatorial logic. Hence, for instance, the particular suitability of the pistils and stamens of the flowering plants. Where animals were concerned, the equivalent might be numbers of feet or teeth (though such discriminants ordered the beasts much less tidily than the sexual system managed the vegetable kingdom). To be sure, there were many who were acutely aware of the shortcomings of such "artificial" systems, and who dilated on the human, limited, and finally

[24] See Scott Atran, *Cognitive Foundations of Natural History: Towards an Anthropology of Science* (Cambridge: Cambridge University Press, 1990), p. 202.

[25] For a somewhat dated chronicle-style early history of comparative anatomy pre-Cuvier, see F. J. Cole, *A History of Comparative Anatomy from Aristotle to the Eighteenth Century* (London: MacMillan, 1944). A considerable number of more recent specialized studies are available for the late eighteenth and early nineteenth centuries, including: Richard W. Burkhardt Jr., *The Spirit of System: Lamarck and Evolutionary Biology* (Cambridge, MA: Harvard University Press, 1977); Larson, *Interpreting Nature;* and Toby Appel, *The Cuvier-Geoffroy Debate: French Biology in the Decades before Darwin* (New York: Oxford University Press, 1987). For the later nineteenth century, see: Mary P. Winsor, *Reading the Shape of Nature: Comparative Zoology at the Agassiz Museum* (Chicago: University of Chicago Press, 1991); and idem, *Starfish, Jellyfish, and the Order of Life: Issues in Nineteenth-Century Science* (New Haven: Yale University Press, 1976). As Winsor herself points out (*Reading the Shape of Nature*, p. xii), the history of systematics is still largely unknown.

[26] The joke, attributed to Edward Forbes, turns up in Huxley's essay "Owen's Position in the History of Anatomical Science," which appears as an appendix to the second volume of: Richard Owen, *The Life of Richard Owen* (London: Murray, 1895). NB: The stereotype tells us as much (if not more) about how the comparative anatomists disparaged their forebears as it does about natural history in the eighteenth century.

merely *convenient* character of these approaches, which clearly failed to cut the world at its joints. But cutting the world at its secret joints, rather than on its manifest dotted lines, was no easy matter, and advocates of more ambitious "natural" systems confronted the daunting task of taking everything into account in each case, or of offering a defensible rationale for electing a particular array of prioritized considerations.

The "joints" of the world were, after all, hidden under the skin of things, and it was by codifying and promulgating a particular method for accessing such internal characteristics that Cuvier achieved his renown as the nineteenth century's leading practitioner of the science of classification. As Daudin puts it in the classic study of the question:

> The resort to dissection, and the complete examination of internal organization—already practiced before Cuvier, but which he did more than anyone else to make into the fundamental and constitutive method of general zoology—would seem to have been the technical factor that gave rise to decisive progress in the properly scientific formulation of zoological groupings.[27]

Cuvier's push into the internal configurations of living creatures precipitated a veritable revolution in the practice of classification.[28] By privileging internal organization Cuvier came to identify a small set of separate "plans" upon which animals appeared to be built; such plans constituted, for him (and for many who followed the multiple volumes of his *Leçons d'Anatomie Comparée* and his synthetic *Règne Animal*), the

[27] Daudin, *Cuvier et Lamarck,* p. iv.

[28] Though close study of this development reveals that it was slower and in many ways more conservative than the secondary literature sometimes suggests. For instance, in their important 1795 revision of the mammals, Cuvier and Geoffroy proposed a threefold division of the class according to the arrangement (and covering) of the appendages (the "mammifères marins" were the first of these three divisions). The rationale for this move did lie in the extent to which these organs of apprehension and sensation were bound up with the whole "lifestyle" (and thus inner anatomy) of the animal, but the authors were aware that their chosen "indicatory character" was not so very far from the old business of counting feet. Indeed, they were at pains to state that attention to hooves and fingernails did not mean they were just reaffirming the old category of the "quadrupeds": "These different degrees of perfection [in the animal world] depend above all on the division more or less pronounced of the fingers, and on their coverings of greater or lesser delicacy. *But be advised that I am not talking about their number;* the number of these parts offers to natural history a characteristic of very little value" (emphasis added). But the very firmness of this protest underscores the proximity of their view to the received wisdom they sought to supersede. It was, at base, a conservative revolution, in keeping, in many ways, with Cuvier's general character. See Etienne Geoffroy and Georges Cuvier, "Mammalogie," *Magasin Encyclopédique* 2 (1795), pp. 152–190, at p. 172. On Cuvier more generally, see Dorinda Outram, *Georges Cuvier: Vocation, Science, and Authority in Post-Revolutionary France* (Manchester: Manchester University Press, 1984).

primary divisions, or *embranchements*, of the animal kingdom; a hierarchy of organ systems determined subsequent groupings. The significance of Cuvier's gambit looms large: the manifest similarities of behavior, habitat, color, voice, taste, size, or external form were, in principle, rendered irrelevant when confronting the problem of organic affinity; what mattered most was, quite possibly, invisible, at least at first glance. A number of scholars have seen the shift in nothing less than world-historical terms. Take Scott Atran, for instance, who goes so far as to assert, concerning Cuvier's forsaking of the ordinary phenomenal domain, that "[p]erhaps for the first time in the history of thought, appearance would no longer constrain the nature of being."[29] Even dismissing this as hyperbole (transubstantiation comes to mind as one of a number of prior claimants to what is, after all, a dubious distinction), the centrality of the comparative anatomical method to nineteenth-century taxonomy—and to the emerging science of "life itself," *biology*—cannot be denied.[30]

Nowhere were the effects of this approach more closely watched in the late eighteenth and early nineteenth centuries than in the shaping of the zoological class into which human beings appeared to fit, that of the *mammifers*, or mammals. And it is worth recalling that the codification of this class hinged on a relatively small number of "problematic cases," of which the cetaceans were probably the most dramatic and, as we shall see, troublesome. Indeed, the shifting of the whales, dolphins, and porpoises from the category of "Pisces" to the class *Mammalia* can be construed as a decisive (and profoundly counterintuitive, when seen in context) move in the bid to make deep anatomy, rather than any constellation of manifest phenomenal characteristics, the dispositive factor in the ordering of nature. Nineteenth-century scientists seized on the point: the distinguished zoologist Sir William Henry Flower, writing in 1900, asserted that a consideration of whales "leads to great generalizations and throws light on far-reaching philosophical speculations," exactly because it taught that "in the endeavor to discover what a creature really is . . . and to what it is related, the general outward appearance affords little clue, and we must go deep below the surface to find out the

---

[29] Atran, *Cognitive Foundations of Natural History*, p. 210.

[30] For an excellent brief overview of these issues, see Paul Farber, *Finding Order in Nature: The Naturalist Tradition from Linnaeus to E. O. Wilson* (Baltimore: Johns Hopkins University Press, 2000). In his recent work on science in early nineteenth-century France, Charles Gillispie has situated the developments in comparative anatomy with respect to the broader trends toward positivism in the period. See Charles Coulston Gillispie, *Science and Polity in France: The Revolutionary and Napoleonic Years* (Princeton: Princeton University Press, 2004), especially chapters 3 and 9.

essential characteristics of its nature."[31] Historians of science have not overlooked the historical significance of this same issue. As Atran notes: "when bats were definitively dropped from the birds, and whales from the fish, so that both could be joined to the mammals, a very profound, even revolutionary, event in systematics thus occurred."[32] As we will see, the disputants in *Maurice v. Judd* understood the connections and sensed the gravity: the implications of a system that put bats and whales together in the same category with a young (White) woman with a baby at her breast—that essential feature of the *mammifers*—received probing and nervous scrutiny in the court.[33]

Just how did "whales" cease to be "fish"? When did this happen? What was at stake? Answering these questions in the case of *James Maurice v. Samuel Judd*—and beyond it—will demand that we consider not merely the cultural ramifications of the science of life in the late eighteenth and early nineteenth centuries, but the emerging structures of that science itself. To begin, let us turn to New York City in 1818.

FROM DOCK TO DOCKET

For all the cetological wranglings of *Maurice v. Judd*, the word "whale" appears nowhere in the original pleadings for the case, which were entered in the manuscript register of the New York Court of Common Pleas on the third Monday of October, 1818.[34] What we find instead is

[31] See William Henry Flower, "Whale," *Encyclopedia Britannica*, 11th edition (Cambridge: Cambridge University Press, 1910–1911), vol. 28, p. 570. The article was written a decade earlier, and appeared in at least one previous edition of *Britannica*.

[32] Atran, *Cognitive Foundations of Natural History*, p. 268.

[33] We now generally hear the word "mammal" without thinking of breasts, but this was by no means the case in the late eighteenth century, when the new category could seem a barbarous (or even perverted) innovation. For an account of the place of gender in the adoption of the term, see "Why Mammals Are Called Mammals," chapter 2 of Londa Schiebinger's *Nature's Body: Gender in the Making of Modern Science* (Boston: Beacon, 1993).

[34] To make a complicated case still harder to trace, the scribe unfortunately recorded the wrong year (1819) in the docket ledger. I would like to thank Bruce Abrams of the Old Records Room of the County Clerk Archives at the New York Supreme Court for helping me secure the pleadings, the minutes, and the jury roll for *Maurice v. Judd*. A general note on the sources for this study: Valuable as the manuscript court records are for confirming the administrative contours of *Maurice v. Judd*, this trial took place in an unfortunate hiatus between the more extensive eighteenth-century court minutes discussed by Richard B. Morris in his *Select Cases of the Mayor's Court of New York City 1674–1784* (Washington, DC: American Historical Association, 1935) and the formalization of court recording, which did not take place until after the judicial reforms in the New York courts in the 1820s. This means that the official court records for the case are sparse on the testimony of witnesses and the proceedings of the case itself (no verbatim transcript of the case appears in either *The New-York City-Hall Recorder* or *The New-*

the trace of a minor commercial transaction that ran afoul of a zealous civil servant:

> James Maurice, plaintiff in this suit, complains of Samuel Judd, defendant in this suit, in custody, &c. of a plea, that he render to the said plaintiff the sum of seventy-five dollars, lawful money of the State of New-York, which to him he owes and from him unjustly detains. For . . . the said defendant, heretofore, to wit, on the first day of July, in the year of our Lord 1818, at the city, and in the county of New-York, and after the 31st day of March, 1818, did buy of one John W. Russell, in the city of New York, three casks of fish oil; the said three casks of fish oil, at the time of the said purchase by the same defendant, not having been gauged, inspected and branded, according to law, contrary to the form of the statue in such a case . . .

Since the statute provided for a fine of twenty-five dollars per uninspected cask, James Maurice, the appointed "inspector of fish oils" in the city of New York, intended to recover from Samuel Judd, proprietor of the New-York Spermaceti Oil & Candle Factory at 52 Broadway (and

---

York Judicial Repository). For this reason, we are obliged to rely in many places on Sampson's published transcript (*IWF*). While this is less than ideal, given his active role in the case, it is less problematic than one might initially think. The text does evidence some suspiciously neat details (favorable to the plaintiff's case), which I will discuss where they are relevant, and there can be little doubt that Sampson gives himself many of the best lines in the case. Even so, it is worth remembering that Sampson was a very active member of a small coterie of early court-affiliated shorthanders and stenographers: he was involved with the publication of a pamphlet that dealt indirectly with shorthand itself (Samuel Woodworth, *Beasts at Law, or, Zoologian Jurisprudence* [New York: J. Harmer and Co., 1811]); and he was immortalized in the first monograph essay in an improbably titled series known as the "Little Visits to the Homes of Eminent Stenographers" (i.e., Charles Currier Beale, *William Sampson, Lawyer and Stenographer* [Boston: n.p., 1907]). He was responsible for more than a dozen published trial records, and there is no evidence I know of that points to any of the participants in these trials challenging his versions of events. Given that a number of those trials, like *Maurice v. Judd*, involved leading civic figures, it seems unlikely that, were his transcripts unacceptable versions of events, no such animadversions would survive, particularly in view of the historical scrutiny which several of his legal exploits have drawn. See: "William Sampson and the Codification Movement," chapter 3 of Maxwell Bloomfield's *American Lawyers in a Changing Society, 1776–1876* (Cambridge, MA: Harvard University Press, 1976); and Walter J. Walsh, "Redefining Radicalism: A Historical Perspective," *George Washington Law Review* 56 (1991), pp. 636–682. Manuscript letters from Sampson in the Carey Papers at the Historical Society of Pennsylvania shed some light on Sampson's financial and professional strategies as a legal author and pamphleteer. Coarsely summarized, they indicate that the business never made him rich. Finally, where *Maurice v. Judd* is concerned, it is worth noting that Sampson identifies the recorder who kept the minutes from which the published transcript of the trial was composed (he was Joseph D. Fay, a lawyer whose name is associated with a number of formally published trial transcripts from the

purveyor of "Spermaceti, wax and tallow mould candles," "winter pressed and summer strained spermaceti and olive oil," as well as an array of lamps, wicks, and lampglasses, at his stand on the old Fly Market), the sum of seventy-five dollars.[35]

Although Judd counter-pled that he "does not owe the said sum of seventy-five dollars," he never contested the assertion that he had purchased casks of spermaceti from Russell that did not bear Mr. Maurice's seal. What Judd contested was that those particular casks demanded Mr. Maurice's statutory attentions (attentions for which, after all, Judd himself was obliged to pay).[36]

At issue was the act passed by the New York State Legislature at Albany on the 31st of March 1818, which stated in relevant part:

> That it shall be the duty of each person appointed by virtue of this act, to provide himself with proper instruments for gauging and inspecting oil, and whenever called on to gauge and inspect any parcel of fish oil, within the place for which he was appointed, it shall also be his duty to inquire diligently, and seek out any parcels of fish oil within his district, and gauge and inspect the same, and brand legibly on the head of each cask he may so gauge and inspect, his own name and the name of the place for which he was appointed; also the whole number of gallons the same shall gauge, and separately from each other the quantity of water, the quantity of sediment, as well as the quantity of pure oil he shall find therein, and shall make, subscribe, and deliver to the owner or holder of such parcel of oil so gauged and inspected, a certificate, exhibiting in separate columns the quantity of each of the aforesaid enumerated ingredients the whole parcel shall contain; for all of which gauging, inspecting, branding, and certifying aforesaid, he shall receive from the

---

period), and that Sampson thanks several of his other colleagues for sharing with him their notes, to aid in his reconstruction of the "fair and full statement of all that was said or acted on this memorable case" (*IWF*, p. vi). For further discussion of trial transcripts in this period, see: Michael Jonathan Millender, "The Transformation of the American Criminal Trial, 1790–1875," Ph.D. dissertation, Princeton University, 1996, chapter 1; and Robert A. Ferguson, *Law and Letters in American Culture* (Cambridge, MA: Harvard University Press, 1984), chapter 3.

[35] See advertisement in *Evening Post* for Saturday, 12 September 1818. NB: There was in fact some question as to whether Maurice was actually the commissioned inspector. It would appear that he was not able to show his commission papers. This technical ploy by the defense was rejected by the judge, and did not affect the verdict.

[36] While merchants were required to pay the inspection fee, they were permitted to pass on a portion of this expense to buyers in the form of a slightly elevated price on retail sales.

owner or holder of the oil so gauged and inspected, twenty cents for each cask, be the same small or large. . . [37]

From the start of the trial itself (on the 30th of December of the same year), the plaintiff's lead lawyer, John Anthon—who along with Sampson, represented the inspector, Maurice—worked to keep the trial on the narrow issue of commercial regulation, and away from the muddy matters of taxonomy: "In a well-ordered community," Anthon announced in his opening remarks, "where an attempt to deviate from the plain path of honest dealing is detected, the legislature interferes, and restrains such wanderings by penal statutes."[38] Because there had been active complaints about the fish oils sold in New York State, the Albany legislature had drafted the act for the inspection of fish oils. And since Judd had manifestly broken that law (by having in his possession uninspected oil), and Maurice had caught him, the case should be an easy one: "The facts in this cause," Anthon asserted, "are comprised in a very narrow compass."

Good to his word, Anthon attempted in the first instance to rest the plaintiff's case after calling only a single witness to the original transaction. Had the casks been conveyed? In New York? After March of 1818? Without marks of inspection? Having established these points to their satisfaction, Maurice's lawyers proposed to rest their case and await the defense. But Judd's attorneys—General Robert Bogardus and William M. Price—replied with a high-stakes gamble, declaring that if this was the totality of the plaintiff's case, then the defense would not call a single witness, and would instead move for a non-suit, alleging that the central question (Was the oil in question *fish* oil?) had not been touched. A *tête-à-tête* at the bench ensued, and Anthon capitulated, agreeing to expand the plaintiff's case to the taxonomic questions—though not without a certain disgust, since, as he put it, he had "supposed it understood that the question was to be tried fully upon its merits, and not like a game of brag."[39]

The comment suggests Anthon knew full well that the floodgates had been opened. His bid to hold the looming classificatory issue at bay had failed: whether "spermaceti oil" was "fish oil"—and thus the vexatious question of whether a whale was a fish—all this was now on the table; and it was to these determinative questions that the next two days, and nearly twenty witnesses, would be addressed. Not only would those

[37] *IWF,* p. 10.
[38] Ibid., p. 2.
[39] Ibid., p. 15.

witnesses leave in the hands of the jury conflicting answers, they would also offer conflicting accounts of who was best positioned to offer answers with authority. As Sampson pointedly put it to Mitchill on the stand later that day: "Doctor, you have mentioned three classes of men, fishermen, artizans, and men of science. There is a much larger class, those who neither fish, manufacture, nor philosophize; have you ever thought it worth while to pay attention to their opinion?"[40]

To tease out the many strands of argument braided into the transcript of *Maurice v. Judd*, I propose to use these four helpful "actor's categories" referenced at this moment in the testimony, and to revisit the central issue of the trial from each of these distinctive perspectives. Was a whale a fish? To make sense of this question we must do as the court did in late December of 1818—we must ask around, consulting those who had a claim to know: those who "philosophize" (the naturalists); those who "fish" (sailors and whalemen); those who "manufacture" (artisans, jobber-merchants, and dealers in whale products like oil and candles); and, finally, as Sampson would have it, that "much larger class" consisting of everyone else. Let us begin with this last group—the most general, "default," category of ordinary English-speaking New Yorkers who had no direct stake in whales or their commercial derivatives. Many of them were in attendance.

[40] Ibid., p. 26.

TWO Common Sense

MANHATTAN AND ITS WHALES

Several comments in the transcripts of *Maurice v. Judd* make it clear that the galleries of the Mayor's Court were very crowded on the last days of the year in 1818. In his closing arguments, for instance, Sampson declared that "it needed no handbills to collect the crowd that has filled this court during the whole trial." The simple rumor of such a "paradox," to be debated by several of the most learned men of the city, "was enough, if the living had no curiosity, to bring the dead out of their graves." That same "paradox" guaranteed the coverage of the trial in a host of newspapers and periodicals in New York and beyond—eventually across much of the young Republic.[1]

The paradox in question was simply the proposition that a whale was *not* a fish. What could be more preposterous? As Anthon explained to the jury: "according to the common understanding and acceptation of men, it would seem that this could hardly admit of a question." And yet, "it has . . . remained for the great lights of the present age to draw it into serious, grave discussion, and to exclude the whale from the family of fish."[2] Nor was this sense of surprise merely the theatrical special pleading of the plaintiff's counsel, who clearly had an interest in presenting his opponent's position as an egregious departure from habit, custom, and verity: published and unpublished responses to the trial support Anthon's assertion that the vast majority of Americans not only assumed that a whale was a fish, but were surprised to learn that the question could be debated. The *National Advocate*, reporting on the "very long and sage trial" on the morning of the first of January of 1819, informed readers that what had emerged from the case was the discovery that learned opinion appeared, on balance, to maintain that whales were *not* fish. The editors had this to say in response: "this is stretching a point of learning to a dangerous length, and all the minute distinctions of genera in the science of ichthiology [*sic*], will not prevent a man of common sense from believing that a whale *is* a fish."[3]

---

[1] *IWF,* p. 66. Other clues include Anthon's allusion to the "crowded audience," p. 2; see also Bogardus on p. 54. I will take up the extent of press coverage of the trial in the conclusion, chapter 7.

[2] Ibid., p. 12.

[3] *National Advocate,* 1 January 1819, p. 2; emphasis in original.

Still more revealing on public opinion is a private letter authored by Henry Meigs—the scion of a distinguished New York political family (and himself a competent amateur naturalist) who was shortly to be elected to represent the second congressional district in the sixteenth Congress. Meigs had followed the trial closely, and writing to his father, a judge in Washington, DC, on the 2nd of January, 1819, Meigs reported on the uproar that attended this sensational case, explaining that Mitchill had testified in court *"that a whale is no fish"*:

> Common sense is quite angry with the Doctor for thrusting his distinctions of mammalia &c down their throats. They don't care a pin whether his blood be cold or hot; mammalia or *Magnolia* he swims & lives in the sea and they swear he is a fish.[4]

That Mitchill's controversial position could be called a "paradox" (literally, *against* the *doxa,* the "teaching") helpfully reminds us that there was indeed a *doxa* on the question, a teaching among Mitchill's contemporaries that established the taxonomic status of the whale. Meigs hinted at this vernacular taxonomy when he characterized the logic of his fellow New Yorkers for his father: everyone knew that the whale "swims & lives in the sea," hence, among the hoi polloi, the whale was a fish, QED. Meigs left out a key part of the syllogism, but that was because he could safely assume his father understood the ordinary English usage of the day. We modern readers may need a reminder: in the late eighteenth and early nineteenth century the word "fish" meant (as the 1817 Philadelphia edition of Samuel Johnson's *Dictionary of the English Language* stated clearly) "an animal existing only in water."[5] At the trial, Anthon would invoke this very reasoning, announcing to the jury that, putting aside the various anatomical and physiological eccentricities of the *cetes,* "the whale remains a fish until naturalists can show him existing on dry land."[6] And the author of the article in the *National Advocate* made the tacit taxonomy on which this classification rested

---

[4] H. Meigs to J. Meigs, 2 January 1819, Meigs Papers, New-York Historical Society; emphasis in original.

[5] "Fish. s. an animal existing only in water." Joseph Hamilton, *Johnson's Dictionary of the English Language in Miniature* (Philadelphia: M. Carey, 1817). The definition in Johnson's 1770 London edition is even more expansive: "An animal that inhabits the water." Samuel Johnson, *A Dictionary of the English Language* (London: Strahan, 1770).

[6] *IWF,* p. 13. The term "cete" (from the Greek, *ketos,* via the Latin, *cetus*—both loosely used to refer to large sea creatures) is no longer used by biologists to refer to the living whales, dolphins, and porpoises, but, in keeping with early-nineteenth-century usage, I will use it in this book where modern readers would expect to find "cetacean" or "Cetacea."

still more explicit: "a whale *is* a fish, for the simple reason that it is not a beast or a bird."[7]

This last aside demands closer scrutiny, since it valuably hints at the principled (and, indeed, authoritative) conception of the "order of nature" that informed popular opinion on the taxonomy of whales. The issue was not that no one understood, for instance, that whales gave birth to live young, or breathed air. In fact, Anthon confessed all such details, and during the course of the trial several of the witnesses—individuals without any scientific training—reflected an awareness of exactly these characteristics. It was rather that such differentia were seen to be secondary or tertiary considerations, and not relevant to establishing where whales belonged in the primary threefold division of animate creation, the division that sorted all creatures into "fish," "beasts," and "birds"—i.e., those that swam in the deep, those that crept on the land, and those that flew in the sky.

The authority for this tripartite taxonomy was nothing less than the first chapter of Genesis, where two verses relevant to the creation, 26 and 28, organized non-human creatures according to the elemental categories of water, earth, and air. In Genesis 1:26, after making man in His image, God gives to humanity its inheritance, declaring: "have dominion over the fish of the sea, and over the fowl of the air, and over the cattle, and over all the earth, and over every creeping thing that creepeth upon the earth." And in Genesis 1:28 the same subdivisions are reinforced: "have dominion over the fish of the sea, over the fowl of the air, and over every living thing that moveth upon the earth." Citing these verses, Anthon explicitly set ordinary language (the language that, for instance, made a *fish*-monger of the peddler of "oysters, crabs, and clams"[8]) and scriptural tradition against the "visionary theories of modern times": "[W]e shall rely on the sacred volume as conclusive," he declared. "From it we learn the great division of all created things, fixed by the deity himself, and which naturalists may mar, but cannot mend, is, the birds of the air, the beasts of the field, and the fish of the sea."[9]

Moreover, by these lights, the whale was not just any fish. Rather, it was, as the Philadelphia *Lady's and Gentleman's Weekly Literary Museum and Musical Magazine* commented in its article on the trial, the very "king of the scaly tribe," the *primogenitus* of the seas.[10] A survey of juvenile

[7] *National Advocate*, 1 January 1819, p. 2; emphasis in original.
[8] *IWF*, p. 56.
[9] Ibid., p. 13.
[10] *Lady's and Gentleman's Weekly Literary Museum and Musical Magazine* 3, no. 18 (22 February 1819), p. 144.

literature and didactic volumes touching on natural history available in
New York before 1819 not only confirms that whales were regularly included among the fish, but also demonstrates that they were (on the
strength of scripture) frequently enshrined as the crowning creature of
that class.[11] Telling in this instance is the pamphlet entitled *The History
of Fish,* published by Samuel Wood and Sons in New York City circa
1816, and available at the "Juvenile Book-Store" on Pearl Street, just a few
blocks north of the Mayor's Court. The frontispiece of this work [Fig-
ure 1] depicts, under the title, a wood engraving of a spouting whale-
creature, and another quotation from Genesis (chapter 1, verse 21) that
would do its duty in *Maurice v. Judd:* "And God created great whales, and
every living creature that moveth." The very first chapter of *The History
of Fish* consisted of an essay on whales (longer than any other chapter in
the book), which detailed their great size and strength, discussed the
commercial products to be derived from them—oil and whalebone—
and recalled the eighteenth-century shore-based enterprise in South and
East Hampton on Long Island that used to pursue them in local
waters.[12] In addition, the text mentioned a near encounter for New York
City itself: "In the year 1775 or 76, came into the sound about twenty
miles from New York, two whales, an old and a young one; the latter was
caught, and taken on shore at Mamaroneck; it was about eighteen or
twenty feet in length."[13] Other juvenile texts available in the same period
also placed the whales at the head of their catalogs of "wonderful fishes,"
and scholars of biblical natural history (a minor genre in the period,
generally consisting of essays on the plants and animals mentioned
in the Bible—ideally suited to pious schoolroom instruction) certainly
affirmed this view.[14] Indeed, most Americans probably first encountered

[11] My aim in this brief section is to address the call made by Sally Gregory Kohlstedt in her
History of Science Society's distinguished lecture of 1989, where she pressed historians of science to delve more deeply into popular conceptions of science in America, particularly in the
early nineteenth century: "we should inquire into the outlook on science taught to children. . . .
we should know what texts, popular treatises, and scholarly monographs sold well, how they
were written, and what they included" (Kohlstedt, "Parlors, Primers, and Public Schooling,"
p. 425). Helpful in this respect is Peter Benes, "To the Curious: Bird and Animal Exhibitions in
New England, 1716–1825," in *New England's Creatures: 1400–1900,* edited by Peter Benes (Boston:
Boston University Press, 1995), pp. 147–163.

[12] "Whalebone" is the common term for the flexible strips of baleen removed from the
mouths of the mysticete whales and used, in the nineteenth century, in a variety of consumer
and industrial products that called for light, flexible, and elastic ribs or springs.

[13] *The History of Fish* (New York: Samuel Wood and Sons, 1816), p. 4.

[14] For other primers, see, for instance, *A History of Wonderful Fishes and Monsters of the Ocean*
(Dublin: Graisberry and Campbell, 1816). Compare, however, William Mavor's *Catechism of
Animated Nature, or, An Easy Introduction to the Animal Kingdom: For the Use of Schools and Fami-*

THE

History of Fish.

And God created great whales, and every
living creature that moveth. Gen. i. 21.

NEW-YORK:

PRINTED AND SOLD BY S. WOOD & SONS,

AT THE JUVENILE BOOK-STORE,

NO. 357, PEARL-STREET.

1816.

FIGURE 1. The King of the Sea: the spouting fish, as depicted on the title
page of a book for sale around the corner from the Mayor's Court. Courtesy
of Beinecke Rare Books and Manuscript Library, Yale University.

the whale in the context of nursery-rhyme natural theology: the ubiqui-
tous *New England Primer* reserved the letter "W" as an occasion to cate-
chize Christian youths in the pervasive power of the Lord's call, using a
couplet that would have been quick to the tongue of every New Yorker
in 1818, "Whales in the sea / God's voice obey."[15]

The reference to whales at Mamaroneck (i.e., in New York waters)
raises the question of what contact the citizens of New York might have
had with the "cetes," other than through their candles (the well-to-do
could afford clean-burning white spermaceti candles, made from a sweet-
smelling oily wax collected from the head cavity and body fat of sperm
whales), umbrella ribs, and corsets (both fashioned from strips of "whalebone," the elastic strips of baleen taken from the mouth of right and
bowhead whales). There was by this time no longer any whaling out of
the city's ports, and only a handful of vessels, sailing irregularly, still
worked from the other shipping centers up the Hudson. While Sag Harbor continued to be an active whaling harbor, it was dwarfed in scale by
Nantucket and New Bedford, and the eastern tip of Long Island was
anyway a long way from Manhattan.[16] Strandings in New Jersey or Long
Island were uncommon enough to merit mentions in the New York City
papers when they occurred, particularly when there was a chance to get
a good view. In 1814 a 22-foot cetacean of some sort blundered up the
Delaware River and was briefly exhibited near Trenton, but that would
have been beyond the reach of the citizens of New York (at least those

---

lies (New York: Samuel Wood and Sons, 1819), which is a question-and-answer format redac-
tion of the 10th edition (1758) of Linnaeus's *Systema Naturae,* and which makes the "cetes" the
seventh order of Mammalia (it is noteworthy, though, that this text was published after the trial
of *Maurice v. Judd*). For a relevant example of biblical natural history, see Thaddeus M. Harris, *The Natural History of the Bible* (Boston: I. Thomas and E. T. Andrews, 1793). Harris (who
was familiar with Linnaeus) not only considered whales fish (p. 167), he appears to have designated the crocodile a "fish" as well.

[15]On this tradition, see Charles F. Heartman, *The New-England Primer Issued Prior to 1830*
(New York: R. R. Bowker, 1934). The alphabets were often illustrated, and little woodcut whales
commonly accompanied the letter "W." John W. Francis, for one, suspected that Mitchill
had a hand in seeing to it that this whale couplet was increasingly replaced by the "equally
sonorous lines," "By Washington / Great deeds were done," though there is little hint, other
than a general patriotism, for why he might have troubled himself with this revision. See John
W. Francis, *Reminiscences of Samuel Latham Mitchill, M.D., LL.D.* (New York: John F. Trow,
1859), p. 11.

[16]One of the witnesses in the case asserted that "he never heard of a whale ship in Albany,
nor had he known of more than three coming into New-York in 15 or 20 years" (*IWF,* p. 15). See
also: Alexander Starbuck, *History of the American Whale Fishery from Its Earliest Inception to the
Year 1876* (Waltham, MA: Published by the Author, 1878); and Richard C. McKay, *South Street:
A Maritime History of New York* (New York: Putnam, 1934).

not interested in a tiresome day-trip).[17] This creature never made the post-mortem journey to Manhattan, but a decade earlier, in the spring of 1804, a still larger whale, taken in the same river and shown at the wharves in Philadelphia, had actually been towed up to New York City by a scrappy entrepreneur and exhibited "from sunrise to 7 o'clock PM" at one of the slips in the East River.[18] Despite being some thirty feet long the carcass did not, in the end, attract the enthusiastic attentions of the populace, largely because by the time it came to be moored (at Ackerly's wharf, not far from Wall Street) it was *more than a month old*, having already been exhibited for several weeks at Philadelphia. The proprietor's defensive advertisements in the New York papers—denying the "report prevailing" that his curiosity was "offensive," and insisting that the only smell was that which "naturally arises from the Blubber"—make clear that the beast was in a sorry state as the weather warmed up; New Yorkers evidently worked harder to stay away than to visit.[19] As we shall see, *Maurice v. Judd* itself hugely stimulated interest in whales and their anatomy in the city after December of 1818, but what about before that? Was there any place in the city that a curious New Yorker could see a whale?

The answer is yes. As of August, 1818, there were in fact two good-sized bits of whale skeleton in the city, and both of them were to be found in the grand new civic emporium of learned culture, the "New-York Institution of Learned and Scientific Establishments" [Plate 1], located on Chambers Street on the north side of City Hall Park, about thirty yards from the Mayor's Court itself (which was held inside City Hall) [Plate 2]. Since I will suggest, in the end, that urban geography matters in the story of *Maurice v. Judd*, it may be worth taking a moment now to look at the period map of lower Manhattan [Plate 3] and the detail of City Hall Park [Plate 4] to see how these buildings were situated, and to get a sense of what the civic heart of the city looked like in those days.

The brainchild of the Princeton-educated (and sometimes bankrupt) merchant, Masonic muckety-muck, and Tammany sagamore, John Pintard, and brought into being by the controversial mayor (and failed presidential candidate) DeWitt Clinton, the New-York Institution emerged in 1816 out of the lobbying of a group of "gentlemen of taste & literature"

---

[17] There are mentions of the incident in the New York papers. For a discussion, see Harry B. Weiss, *Whaling in New Jersey* (Trenton: New Jersey Agricultural Society, 1974), pp. 105–110.

[18] *Utica Patriot*, 14 May 1804, p. 3.

[19] *New-York Herald*, 21 April 1804, p. 2.

preoccupied with the city's unhappy reputation as an exclusively mercantile (perhaps even mercenary) center of getting and spending, a place without (in contrast to Boston and Philadelphia) much regard for arts, letters, or science.[20] As the Common Council of the city put it in that year, "The Citizens of New York have too long been stigmatized as phlegmatic, money making & plodding—Our Sister Cities deny we possess any taste for the sciences."[21] The completion of a new almshouse at Bellevue, and the changing physical and social setting of City Hall Park (which now featured the City's architectural pride, Joseph Francois Mangin and John McComb Jr.'s new City Hall building), had left the "Old Almshouse" at the top of the park vacant, and it was this structure that DeWitt Clinton secured as the free home for a handful of New York's learned societies and institutions: his administration would display its commitment to a high-minded patronage of learning by means of actual bricks and mortar, and City Hall Park would become an august citadel of knowledge and power, with Mangin and McComb's white marble wedding cake flanked by a broad palace of arts, letters, and science.[22] Though satirists jeered the parade of worthies establishing themselves in the halls once thronged with ragpickers and orphans, the lure of civic subsidies nevertheless prompted a run on the rent-free space, and behind-the-scenes jockeying determined the small pool of lucky leaseholders, which included, when the dust settled by the summer of 1817: the New-York Historical Society, the American Academy of Fine Arts, the Literary and Philosophical Society, the New-York Society Library, John Griscom's Chemistry Laboratory, and the Lyceum of Natural History; John Scudder's "American Museum" would be added to the roster shortly thereafter.[23]

[20] On the founding of the Institution see Bender, *New York Intellect*, pp. 62–64; Bender is here working primarily from the valuable and detailed dissertation by Nodyne (see Nodyne, "The Role of De Witt Clinton").

[21] See Bender, *New York Intellect*, p. 64, and Edwin G. Burrows and Mike Wallace, *Gotham: A History of New York City to 1898* (New York: Oxford University Press, 1999), p. 467. See also "On the Means of Education and the Scientific Institutions in New York," *Analectic Magazine* 13 (1819), pp. 452–459, at p. 459: "From the above remarks it will be seen that New York possesses very many of the sources, which will give rise to public improvement in knowledge, and is rapidly acquiring more; and if the progress be continued in the same ratio, she may soon rank as high in the intellectual scale as she does in the commercial."

[22] For a discussion of changing ideals and experiences of urban space in Manhattan in these years, see Wyn Kelley, *Melville's City: Literary and Urban Form in Nineteenth-Century New York* (Cambridge: Cambridge University Press, 1996), particularly chapter 1.

[23] Note that Bender cites an erroneous date for the Lyceum's move (he gives 1820). The manuscript minute books of the Lyceum itself, now held by its surviving successor institution,

This consolidation of New York's coterie of older learned societies (which had much overlap in their membership rolls), together with a new nascent quasi-professional society for naturalists (the Lyceum), a performing chemist-apothecary dedicated to public instruction (Griscom), and a commercial enterprise for the edification and entertainment of the people (Scudder's Museum), all within a stone's throw of the city's political and legal nucleus, represents a significant shift in the character, visibility, and cultural authority of science in New York in this period. DeWitt Clinton, thinking in terms of posterity, invoked England's Royal Society itself, and by the 2nd of November of 1818, David Hosack (the London- and Edinburgh-educated botanist, who had studied with James E. Smith, the president of the Linnaean Society of London, and who had established the Elgin Botanical Garden in New York City upon his return, taking over Mitchill's professorship at Columbia College) could announce that, thanks to the "liberal Corporation," the New-York Institution had helped elevate the city of New York's literary, medical, and scientific character "to that rank which in other respects she has long enjoyed."[24]

For Hosack, moreover, the availability of natural history specimens was a key feature of New York's intellectual maturity:

> In the Cabinet of Natural History belonging to this College [the College of Physicians and Surgeons], in connexion with those attached to the New-York Historical Society and Lyceum of Natural History, and the valuable collection that has been amassed by that indefatigable and skilful collector and preserver of the productions of nature, the proprietor of the New-York Museum, <* Mr. John Scudder> the student is presented, under the guidance of Dr. Mitchill, our learned professor of Natural History, with a view of the most important objects that can arrest your attention in any of the branches of that useful science.[25]

---

the New York Academy of Sciences, confirm that the Lyceum took over space originally conferred to J. G. Swift's U.S. Military and Philosophical Society by April of 1817 (the Lyceum's first meeting in the new space was on the 21st of that month). It is a little more difficult to say for certain when the Lyceum's collection of *naturalia* was installed in the new space, but a description of a Lyceum meeting in the Institution published in the *Columbian* on 28 July 1818 makes it clear that the participants had their collections to hand, so this establishes a *terminus a quo* for the cabinet.

[24] David Hosack, "Progress of Medical Science in New-York," *American Monthly Magazine and Critical Review* 4 (1818), pp. 114–116.

[25] Ibid., p. 114. The asterisked text represents a footnote in the original.

Two of those collections featured partial whale skeletons: the cabinet of the Lyceum of Natural History (which owned the skull and jaw of a large right whale), and Scudder's Museum (which held the jaw from another large baleen whale, almost certainly also from a right). These were very different institutions, and very different collections. In a sense, we might say that one of these whale specimens belonged to "those who philosophize," and the other belonged to "everyone else."

The first and more complete set of bones was part of the private cabinet of the newest and most dynamic society for the study of nature in New York City, the Lyceum of Natural History, which received for its collection on Monday, the 9th of June, 1817, "the head and jaws of a whale, *balena mysteicetus* . . . from Mr Martin Van Buren of Flatbush."[26] The Lyceum, founded in early 1817, represented an upstart private society of dedicated naturalists, who had come together in the hopes of emulating Philadelphia's Academy of Natural Sciences.[27] Just as this latter institution had formed itself as an alternative to what many of its members saw as the fusty collective of patrician generalists that dominated Philadelphia's learned culture (the American Philosophical Society), so, too, the animating spirits behind the Lyceum were taken with the idea of a society for natural science more focused than either the Literary and Philosophical Society or the Historical Society, both of which still worked to represent all the relevant branches of polite learning, from numismatics to malacology, mathematics to philology, theater to meteorology. By contrast, the members of the new Lyceum were expected to aspire to mastery in some domain of the specifically *natural* sciences: the bar for membership would be set to reflect these higher expectations, and all members would be called on to present papers in their areas of expertise, with the hope that these papers would yield publishable contributions in natural history (taken to include zoology, botany, geology, and

[26] Minutes of Lyceum of Natural History, 9 June 1817, collection of the New York Academy of Sciences. A note on nomenclature: Readers familiar with the modern Latin binomial designations for mysticete whales may be confused that I have identified Van Buren's specimen as a right whale (genus *Eubalaena*) when Mitchill and his colleagues have explicitly called it a *Balaena mysticetus*, which now refers exclusively to the bowhead. As I discuss below (chapter 4, footnote 95), the differentiation among the various members of what we now know as the family *Balaenidae* was not common before 1850. Since the Lyceum's whale was secured well to the south of the known range for bowhead whales in modern times, there is every reason to think that the creature in question was what we would now call a *Eubalaena glacialis* (Müller, 1776), a North Atlantic right whale. The same goes for Scudder's whale, which also got called a *Balaena mysticetus* at various times.

[27] On the Academy of Natural Sciences, see Simon Baatz, "Philadelphia Patronage: The Institutional Structure of Natural History in the New Republic, 1800–1833," *Journal of the Early American Republic* 8, no. 2 (1988), pp. 111–138.

a number of related sciences of the earth, sea, and atmosphere).[28] The Lyceum had a small but fast-growing cabinet of natural history specimens, and the records of the early society make clear that the maintenance and expansion of the cabinet were major preoccupations of the members. Within a decade the Lyceum would have amassed the most extensive private teaching collection in the city, buying out all the *naturalia* in the cabinet of the Historical Society (which was increasingly focused on antiquities) and arguably holding the best collection of fish and reptiles in the country.[29] While most of the original collection (including the whale) was destroyed in a fire in 1866, the Lyceum's renewed holdings would go on to form the seed objects in what is now known as the American Museum of Natural History.

In the next chapter, which will take up the status of the whale among "those who philosophize," I will return to the Lyceum, its teaching cabinet, and its whale skeleton. For now we need to focus on the bits of cetacean that could be seen by anyone in possession of a quarter (the equivalent today of, say, a movie ticket) and the inclination to visit the public face of the city's new Institution of Learned and Scientific Establishments.[30]

In contrast to the private society cabinet of the Lyceum, Scudder's American Museum, which took up the galleries on the western end of the Old Almshouse building [Figure 2], was New York's version of the Philadelphia and Baltimore institutions made famous by the Peale family.[31] Vanguard establishments in what Joel Orosz has called the American

[28] There is a large literature on the Lyceum and its relationship to the other learned societies of the period. I base this paragraph on a reading of the first three decades of manuscript minute books for all three societies (the Lyceum, the Historical Society, and the Literary and Philosophical Society), as well as on the following secondary sources: Ralph S. Bates, *Scientific Societies in the United States* (Cambridge, MA: MIT Press, 1965); Nodyne, "The Founding of the Lyceum of Natural History"; Baatz, "Knowledge, Culture, and Science"; Hindle, "Learned Society in New York"; and Harris, "New York's First Scientific Body."

[29] See Herman L. Fairchild, *History of the New York Academy of Sciences* (New York: The Author, 1887), pp. 101–103.

[30] In 1818 twenty-five cents would buy a sermon pamphlet (for the spiritually inclined) or a satirical quarterly (for the irreverent); it was the cost of admission to many outdoor entertainments like fairs or musical performances. Half-price tickets were often available for children.

[31] For a sense of the development of these institutions, and their changing profile 1786–1850, see William T. Alderson, ed., *Mermaids, Mummies, and Mastodons: The Emergence of the American Museum* (Washington, DC: Association of American Museums, 1992). Kohlstedt's chapter, "Entrepreneurs and Intellectuals: Natural History in Early American Museums," is particularly valuable. See also: Lillian B. Miller and David C. Ward, eds., *New Perspectives on Charles Willson Peale* (Pittsburgh: University of Pittsburgh Press and the Smithsonian Institution, 1991); and Toby Appel, "Science, Popular Culture, and Profit: Peale's Philadelphia Museum," *Journal of the Society for the Bibliography of Natural History* 9 (1980), pp. 619–634.

FIGURE 2. Scudder's American Museum: a view of the northwest corner of City Hall Park, looking southeast; Scudder's establishment had signage on the western end of the New-York Institution (City Hall itself is the building on the right). Courtesy of the Picture Collection of the Branch Libraries, New York Public Library; Astor, Lenox, and Tilden Foundations.

"didactic enlightenment" of the period 1800–1820, these public collections occupied an unstable cultural niche somewhere between (on the one hand) the pedagogical ideal of enlightening improvement for all worthy citizens of the Republic, and (on the other) the commercial imperatives of mass entertainment. The trajectory of the objects in the Scudder collection nicely traces the arc between these poles, since the cabinet started out as the museum of the philanthropic Tammany Society in the 1790s, and by the 1840s the collection had been swept under the flamboyant and expanding reach of the "Prince of Humbugs," P. T. Barnum. The years during which the collection resided in the New-York Institution, 1817–1830 (and particularly the early years, before the death of the animating spirit of the collection, John Scudder Sr., in 1821), saw the American Museum's most successful marriage of public instruction and commercial viability.

Scudder had learned the trade of managing such an enterprise from the painter-*cum*-historian-*cum*-showman Edward Savage, whose "City Museum" at 80 Greenwich Street represented merely the last in a string of short-lived panoramas and galleries from Boston to Philadelphia in

which Savage had been involved from 1794 to 1809, when he sold out to
Scudder, his most able apprentice. The new proprietor, who had always
displayed a marked preference for natural history over the beaux-arts
taste of his boss, had by that point built up a small collection of his own
during a hiatus in his employment for Savage, and thus by March of 1810
when Scudder opened his "New American Museum" at 21 Chatham
Street (just to the east of City Hall Park) he had already begun to beef
up Savage's core collection of naval battle tableaux and unusual sculpture
with a set of natural historical objects.

The whale jaw was perhaps his greatest commercial success of all.
Hacked from an animal taken on Long Island's South Shore in the
spring of 1814, and transported by coasting vessel and cart to a choice spot
on Broadway just above City Hall Park, this massive archway of bone and
flesh weighed nearly 4,000 pounds, and a dozen men could stand beneath its six-foot curtains of baleen. It arrived on the 25th of March, and
the newspapers were instantly plastered with advertisements exhorting
curious New Yorkers to "seize the present opportunity" and avail themselves (for the rock-bottom fee of twenty-five cents) of the chance to see
"the jaw of this mammoth of the deep," a wonder "beautiful beyond description."[32] By the next week a festival atmosphere prevailed, and Scudder, never afraid to gild the lily, had engaged his favorite band—a group
of (purported) Italians known as "The Pandean Minstrels" (who often
jazzed-up special events at the museum)—to perform evening concerts
under the whale jaw. Who could resist the "delightful melody on their
Reed Pipes, such as were used in antient [sic] times by Pan, the son of
Mercury and God of Shepherds, in the fields of Arcadia"—particularly
when these bacchantic strains echoed under the arches of the Leviathan?
As one advertisement for the event proclaimed: "Perhaps a scene so
natural, so beautiful, so musical and new in its appearance was never before seen, heard of, or even thought of, in the old or new world."[33] Fair
enough.[34]

Scudder packed them in on Broadway: by the 4th of April the papers

[32] *Evening Post*, 26 March 1814, p. 3.

[33] This and above quotes from *Evening Post*, 2 April 1814, p. 3.

[34] Though it should be said that the Old World knew plenty about impromptu festivals on
the occasion of a whale stranding. For instance, a host of Dutch prints dating back to the sixteenth century (many of them departing from an original drawing by Hendrik Goltzius) depict
carnival gatherings of curious gawkers perusing a cetacean on the strand. There were similar traditions in other European countries. For a sense of some of this, see Elizabeth Ingalls, *Whaling
Prints in the Francis B. Lothrop Collection* (Salem, MA: The Peabody Museum of Salem, 1987).

were declaring the whale jaw the most extraordinary thing to happen in the city since the Revolution.[35] And Scudder milked it for all it was worth: nearly a month later he was still flogging the jaw (which was about to go on the road, for displays in Albany and points farther upstate), though his promise to potential customers—that "owing to the attention it has received from the proprietor, it is perfectly as sweet and destitute of the least disagreeable smell as the day on which it arrived"—suggests his bandstand was getting a little gray around the proverbial gills (*proverbial* gills) by late April.[36] By the end of the summer the jaw—now stripped down to bone and baleen and cleaned up—was back in the city, and (still touted as "the greatest curiosity ever offered to the public") installed in the American Museum on Chatham Street.

Such carnival promotions were the heart of the business, but during these early years of the American Museum Scudder also sought out connections among the more visible naturalists in the city—including Samuel L. Mitchill himself—and made the holdings in his public gallery available as a teaching collection, positioning his emporium on the cusp of university culture. An anonymous author (Scudder himself?), writing in the *Commercial Advertiser* shortly after the opening of the collection, emphasized the value of the specimens for public instruction in natural history: "The selection, preservation and arrangement of the various subjects of Natural History, do infinite credit to the taste, skill and judgement of Mr Scudder. . . . Parents will find this Museum an instructive school to teach their Children to behold and admire the marvelous works of creation."[37] It would seem that Mitchill agreed, since by 1811 he was using the collection to supplement his teaching at the breakaway institution of higher education that would soon be named the College of Physicians and Surgeons. A student who had attended a cycle of Mitchill's lectures in natural history wrote an abstract of the course for the *American Medical and Philosophical Register* in that year, and noted:

> While he [Mitchill] was occupied in tracing the generic and specific characters of animals, he made adjournments day after day, from the college to the museum of Mr. Scudder, and there, in the midst of animal specimens prepared by that indefatigable

[35] "Natural History," *Columbian*, 4 April 1814, p. 3.

[36] *Columbian*, 28 April 1814, p. 2.

[37] Cited in Joel J. Orosz, *Curators and Culture: The Museum Movement in America, 1740–1870* (Tuscaloosa: University of Alabama Press, 1990), p. 75. Periodic plugs for the institution turned up in the *Commercial Advertiser*, suggesting Scudder had an "in" there. See, for instance, pieces from 2 May 1812, 9 December 1819, etc.

collector, with the most exquisite touches of art, he gave practical lessons on the several classes of creatures, to an audience that entered fully into the spirit of the business.[38]

The famous whale jaw itself had drawn Mitchill's attention the moment it landed on the docks: A young Philadelphia Quaker with an interest in natural history, Reuben Haines, visited New York several times in the early teens, and, via a network of Quaker grandees, made Mitchill's personal acquaintance. In addition to attending Hosack's botany lectures at the Elgin Botanical Gardens, and sitting in on Mitchill's introduction to natural history at the new college, Haines had the honor of some personal tuition in marine natural history with Mitchill himself, peripatetic instruction that involved visits to Mitchill's preferred sites for natural history. As Haines wrote a Philadelphia friend on the 5th of April, 1814 (in the thick of the whale jaw craze on Broadway):

> But above all I have had the honour of standing in the Jaw of the Great Whale (*Balaena Mysticetus*) with the learned Dr. Saml L Mitchell [*sic*]—besides receiving a *private* lecture from him in the *New York Fish market* . . .[39]

In 1816, by dint of his strong allies and reputation as a civic-minded entrepreneur, Scudder succeeded in securing the largest space in the new New-York Institution (a room of 96 by 41 feet, with a ceiling high enough to admit a gallery, and glass cases running the length of the walls), and by the 2nd of July, 1817 he had moved the collection over from neighboring Chatham Street, and was ready for a gala opening for his patrons and "those scientific gentlemen who had taken a particular interest in his success"—a private thank you party to be followed by a public *vernissage*.[40] An anonymous letter (again, quite possibly a plant) to the editors of the first volume of the *American Monthly Magazine and Critical Review* had nothing but praise for this new beacon in the city's cultural landscape:

> The brilliant display made on this occasion, gave an opportunity
> for many to admire the taste of Mr. Scudder (the proprietor) in

---

[38] A Correspondent, "Cultivation of Natural History in the University College of New York," *American Medical and Philosophical Register* 2 (1811), pp. 154–163, at p. 161.

[39] Haines to James Pemberton Parke, 5 April 1814, folder 136, box 15, series ii, Wyck Collection, Papers of Reuben Haines III, American Philosophical Society Manuscripts Library; emphases in original. This letter is misdated as 1813 (by Haines himself), but Parke helpfully noted the date he *received* the letter, 7 April 1814, in an annotation that underlines Haines's error.

[40] The dimensions and layout of Scudder's space are given in the *Commercial Advertiser*, 9 December 1819, p. 2. See also Orosz, *Curators and Culture*, p. 77.

the disposition of his natural curiosities, and the elegant manner in which he has prepared and preserved them, and varied their natural attitudes to give the strongest impressions, and produce the most lasting effect upon the beholder. . . . It was the opinion of several gentlemen present [at] the first exhibition, that neither London nor Paris, which they had visited, possessed specimens in such high state of preservation; and that as he already excelled in the preparation, he would soon exceed in the number of his Museum, any similar establishment.[41]

By October of that year Scudder had pressed forward with a contribution to the broad pedagogical program of the New-York Institution. Responding to the concern that the majority of the lectures given at the Institution were presented within the meetings of the various learned societies themselves, and thus "heard only by the scientific gentlemen who compose these societies," Scudder announced that for the "gratification of our citizens" and pursuant to the aim of his museum—"combining amusement with instruction"—he would engage a "professional gentleman to deliver a course of Lectures on Natural History, in the Museum," two or three times a week through the academic season (beginning in October). These lectures—to be held in the upstairs rooms of the American Museum, which would be "fitted up for the accommodation of attendants"—would "embrace a general view of the animal creation, and the specimens in the Museum will be used in illustration of the subjects under discussion."[42]

This public instruction was wholly consistent with John Pintard's founding vision for the New-York Institution. In a letter to his daughter about this brick and mortar "child of my old age" (as he called it) Pintard wrote, shortly before the opening, that he expected that the Institution would sponsor "courses of Lectures in Chemistry, Minerology, Zoology & Botany."[43] Whether such a sequence of public lectures ever actually took place is not clear. What is certain, however, is that, even as he continued to hold the approbation of the city's emerging scientific community, Scudder assiduously attended to his responsibilities as a purveyor of public entertainment. He engaged various musicians to play on the balcony of the Institution on pleasant evenings, and built a reputation for the museum as a focal point for civic pageantry, decorating the

---

[41] "New-York Institution," *American Monthly Magazine and Critical Review* 1 (1817), pp. 271–273, at p. 271.

[42] *American Monthly Magazine and Critical Review* 1 (1817), p. 455.

[43] John Pintard, *Letters from John Pintard* (New York: New-York Historical Society, 1940), vol. 1, p. 60, letter of 15 April 1817.

establishment with emblazoned patriotic dioramas on special occasions like the 4th of July, and advertising in the newspapers in the holiday season that "The Museum will be Brilliantly Illuminated on Christmas and New Year's evenings, and the Museum Band will perform some of the most appropriate and pleasing airs."[44] These same circulars did not belabor the museum's holdings in natural history, but did draw attention to the "Grand Cosmorama" of Vesuvius and Naples, as well as an excellent marble likeness of sleeping Cupid, "executed by a grand master in Italy."

Pintard himself was pressed to acknowledge the tension between the American Museum's growing popular appeal as a destination for promenades, rendezvous, and civic mingling, and its purported goal of public instruction. After visiting Scudder's on the 4th of July in 1819, he wrote to his daughter that "His museum was thronged":

> But what to me was the most imposing spectacle, was the multitude Males & Females who thronged and filled all the park between his porch and Broadway. The moon, nearly at full, shone resplendent & the blaze of the light from the illumination [i.e., Scudder's brightly lit façade] thrown on several thousand faces, all, as in a theatre, looking towards the Museum, presented a most interesting sight. As to the interior, the multitude came to see the 4th of July, for the crowd was so great, & the heat so insufferable, that no gratification could arise from viewing his elegant preparations.[45]

The music, however, was "very fine."

Charles Willson Peale, no stranger to the delicate matter of holding public entertainment together with moral and natural philosophy, visited Scudder's Museum shortly before it opened, both to scope out a rival, and as part of a mooted Peale-Scudder joint enterprise (which Pintard had tried to broker). While Peale appreciated the collection, he did note in his diary that a number of the animals had, to his eye, been mounted for sensational display rather than for scientific instruction: they had been configured to "excite [the] admiration of those unacquainted with proper attitudes" of such beasts, and a surfeit of shellac and paintbox colors made for "beauty of plumage and slickness of skins," but not necessarily accuracy of form or appearance.[46] There was a bit too much nudity in the print room, too, for Peale's Philadelphia taste.

[44] *Columbian,* 31 December 1818.
[45] Pintard, *Letters,* vol. 1, p. 203, letter of 4 July 1819.
[46] The point is made by Orosz, *Curators and Culture,* p. 79. See also Charles Willson Peale, *The Collected Papers of Charles Willson Peale and his Family, 1735–1885* (Millwood, NY: KTO Microform, 1980), microfiche II B, card 22, diary no. 22, circa 1 June, 4 June, and 10 June 1817.

Because I am interested in recovering the vernacular natural history reflected in Scudder's Museum, and intend to use his inventory and arrangement for clues to what museum-goers in New York in 1818 would have learned about the order of nature (in general) and the place of whales in that order (in particular), we need to consider for a moment just who went to the American Museum. Was it a place of resort for "everyone"? Can it offer us insight into the understanding of "ordinary citizens" like those on the jury of *Maurice v. Judd?*

By far the most detailed effort to examine the social profile of the visitors to American public cabinets in this period is David Brigham's essay on "Social Class and Participation in Peale's Philadelphia Museum," which makes use of extensive subscription rosters to sketch the composition of the most dedicated patrons of that collection. Extending those observations to Scudder's Museum is by no means without its pitfalls, but in lieu of equivalent records there is little alternative. In sum, then, Brigham shows that "in contemporary terms, the upper sort were vastly overrepresented among Peale's subscribers, while the middling sort were underrepresented and the lower sort were absent."[47] This is to say that the majority of those who bought full-year passes for admission hailed from the "high non-manual" occupations (successful merchants, doctors and attorneys, appointed officials of government), though "low non-manual" occupations (like shopkeepers, clerks, and small manufacturers) were also well represented, and "high manual" tradesmen (like leather workers, bookbinders and printers, and apothecaries) amounted to more than ten percent of the total group. This breakdown represents *sub-scribers,* that is, those making a significant commitment to the museum (by 1819, a subscription cost $10), so it is safe to assume that a similar survey of ordinary day-ticket admissions would reflect a considerably broader segment of the population. At the same time, the exclusion of the "lower sort" did represent an explicit aim of the establishment, which traded on aspirations of moral improvement and social grace. The cost of an ordinary admission ticket was low, but Peale wrote in 1824 that "the doors of the Museum have ever been closed against the profligate and the indecent, [and] it has been preserved as a place where the virtuous and refined of society could meet to enjoy such pleasures as can be tasted by the virtuous and refined alone."[48] Such declarations were perhaps more op-

[47] David R. Brigham, "Social Class and Participation in Peale's Philadelphia Museum," in Alderson, *Mermaids, Mummies, and Mastodons,* p. 86.

[48] Single tickets were twenty-five cents for general admission, but special shows (for instance the celebrated mastodon) could demand an additional fee of fifty cents or more. Orosz,

tative than descriptive, particularly in light of Peale's perpetual struggles to maintain the gracious appointments of his public parlor: by 1809 he was obliged to install signage requesting that visitors refrain from plinking at the displays, and in the same year he complained to his son that people not only smudged the glass of the cases and sometimes stood on the leather-covered benches to peer into the higher shelves, but that some even went so far as to scrawl graffiti on the plaster casts, and scratch their names into the bases of the natural history specimens.[49] The sparser evidence on the character of Scudder's establishment suggests a similar institution, one that reached a similar community. It was a place that would certainly have been within the cultural horizon of the more or less reputable citizens selected for the jury in *Maurice v. Judd*—men like the successful brewer William Wilmerding, the hardware dealer Isaac Underhill, the registered auctioneer and dry-goods merchant William S. Hick, or the demimonde lawyer George Niven.[50]

What would they have learned about the natural world from such a visit? The best evidence for the contents and organization of the American Museum is to be found in a unique folio broadsheet printed in 1819 and surviving in the collection of the New-York Historical Society, entitled, "Abridged catalogue of the Principal Natural and Artificial Curiosities, now in the American Museum." This inventory arranged the contents of the natural history collection by type, after the manner of a taxonomy, informing the prospective visitor that the museum held, for instance: "112 cases of birds, containing 676 from different parts of the globe," including 44 hummingbirds; "46 cases of Quadrupeds, containing 113 animals," including "2 leopard seals taken in a net opposite the city"; along with considerable holdings of "fish," "reptiles," and "insects."[51]

Curators and Culture, p. 84. The pursuit of the mastodon in North America is a choice tale of sensation and science in the early Republic. See Paul Semonin, *American Monster: How the Nation's First Prehistoric Creature Became a Symbol of National Identity* (New York: New York University Press, 2000). For a complementary perspective, see Claudine Cohen, *The Fate of the Mammoth: Fossils, Myth, and History* (Chicago: Universtiy of Chicago Press, 2002).

[49] Orosz, *Curators and Culture*, p. 87.

[50] Biographical information on nearly all the jurors can be gleaned from period newspapers. See, for instance: *Spectator*, 9 April 1819, p. 1 (for a mention of Hick's commission as an auctioneer); the classified advertisement in the *Commercial Advertiser*, 21 May 1816, p. 1 (for Underhill's wares); or *Evening Post*, 8 January 1821, p. 1 (for the sale of Wilmerding's establishment). In general, jurors in this period were drawn from a relatively small pool of reasonably well-to-do men.

[51] This is almost certainly not the animal we now know as the leopard seal (*Hydrurga leptonyx* [Blainville, 1820]), which is exclusively Antarctic. American sealers had surely encountered these top predators in the deep southern ocean, but it seems unlikely they would have

The "quadruped" entry signals that Scudder organized his zoology according to traditional classifications, since the class of "four-footed beasts"—while retained, for instance, in the work of Buffon and his many students and pirates—had long been losing its standing as a meaningful grouping among classifiers who followed Linnaeus's tenth edition of the *Systema Naturae* of 1758, and it had slipped quite completely out of formal taxonomy as the techniques of comparative anatomy gained ascendancy in the early years of the nineteenth century. That Scudder put a seal among the "quadrupeds" indicates that he was not unaware that actual four-footedness was no longer a prerequisite for this category. Seals, otters, and walruses had been numbered among the "amphibia" in many eighteenth-century taxonomies, and Scudder's sense that they belonged with his goats and wolves reflected the way that the term "quadruped" had—in the suburbs of university learning, in the mid-Atlantic regions of the young Republic, by 1819—acquired something of the sense that advocates of the "new philosophy" gave to their preferred Linnaean term, "Mammalia."

Turning to Scudder's listing of fish, however, it becomes clear that he had by no means wholly reconfigured his groupings along these new lines. The first three listed "fish" were given as follows:

1) The whale killer, or great Black Fish of the South Seas, was taken off the Island of Nantucket, and when living weighed 18 cwt. Measures 7 feet in circumference and 12 in length.

2) Bottle nose porpoise of the South Seas, was taken in Long Island Sound, December, 1812.

3) The Herring Hog Porpoise of North America.

which Scudder would surely have known were all gill-less creatures that spouted.[52] The lesser fish—like "dolphins" (meaning here *Coryphaena hip-*

---

bothered to bring one all the way back for Scudder. "Leopard" as a term designating spotted coloration has long been so common that it is more probable that Scudder simply means a seal with spots (perhaps the ringed seal, *Pusa hispida* [Schreber, 1775]?).

[52] The first was presumably a killer whale, the last probably a harbor porpoise. Mitchill not only took notes on these specimens, but also quibbled about Scudder's identifications. In the Issacher Cozzens Portfolio Print Collection at the New-York Historical Society (an assortment of sketches and prints that includes a considerable amount of original Mitchill material) there appears a small pencil sketch of a cetacean (PR 145, #67, center of bottom row) that is labeled in Mitchill's hand "Scudder's mus. 15 feet long, D. globiceps? called bottlenose there," which suggests Mitchill thought what Scudder called a bottlenose was more likely a pilot whale. On the Cozzens Collection, see Jenny Gotwals, "Portfolio of a Curious Mind: The Issachar Cozzens Collection," *New-York Journal of American History* 65, no. 4 (2004), pp. 48–57.

purus [Linnaeus, 1758], now referred to most often as dorado, or mahi mahi) and extending even to squid—followed in the list. In addition, at the bottom of the section Scudder alludes to generous holdings of "crustaceous fish" including a large lobster, a South American "crawfish," and "the great land crab from the Spanish Main."

This configuration of "cetes" with crabs and the "sepia" nicely captures the ample signification of "fish" in the period: in the institution with the best claim to serve as the city's popular temple for natural history instruction, "fish" meant, grossly, "creatures from the water."

Not that it was quite that simple. Creatures from the water with four legs did *not* rank with the fish: turtles (including a monster 1,000-pound sea turtle taken by a New York pilot boat just outside of the entrance to the harbor), for instance, got a separate listing; they were not considered "quadrupeds" proper, but nor were they "fish" (or, for that matter, "reptiles," a designation reserved for things that slithered on their bellies, like snakes).[53]

Still more telling, perhaps, where *Maurice v. Judd* is concerned, is how Scudder listed his vaunted whale jaw specimen: he placed this object— which had once been the talk of the town—not under the "fish" at all, but rather under the catchall heading of "Miscellaneous," a category in which he lumped together both natural and artificial curiosities. Hence, together with the whale jaw we discover an "Indian Mummy, dug out of a saltpetre cave in Kentucky," followed by the "teeth of a mammoth" and (less alluring, perhaps) a group of large hairballs, including three "taken out of the bowels of a hog owned by J. Deitz in Spring Street, wt. 10 lbs."[54]

[53] Compare, however, a slightly later vernacular taxonomy (Hardie's enumeration of the creatures for sale in the New York markets), where several species of fresh- and saltwater turtles are listed under the category "Amphibious." James Hardie, *The Description of the City of New-York* (New York: Samuel Marks, 1827), p. 184.

[54] It is interesting to notice that here Scudder listed his jaw as that of a "scrag whale," not a *Balaena mysticetus*. Whaling buffs may wish to meditate on this, since it is certainly odd. The term "Scrag" appears in Paul Dudley, "An Essay upon the Natural History of Whales" (*Philosophical Transactions of the Royal Society* 33 [1725], pp. 256–269), and is now generally thought to correspond to what is today called the "gray whale" (*Eschrichtius robustus* [Lilljeborg, 1861]). It should be emphasized that the retrojection of species names is a very difficult matter, and it seems that the term "scrag" was used quite loosely in this period. Did Scudder in 1819 perhaps have *another* whale jaw, one from what we would now consider a gray whale? It seems impossible: Scudder lists the jaw in question as "containing 300 pieces of bone," but this is considerably more than the top end of the range found in *Eschrichtius*; moreover, most evidence suggests that in historical times grays have been an exclusively Pacific species, and it is quite certain they were hunted almost not at all by Anglo-American whalers in that region before the middle of the nineteenth century (but compare F. C. Fraser, "An Early 17th-Century Record of

A number of explanations might be offered for this seemingly odd grouping. Perhaps the incompleteness of the whale specimen (it was just a jaw after all) meant it did not belong in the fish listing; or perhaps the presence of the whalebone in the jaw made the specimen more salient as the raw material for familiar manufactured commodities than as a proper item of natural history (which might explain how the jaw could be placed beside a "miniature model of the palace of St. Cloud with 144 moving figures, built entirely from beef bones"). What is certain, however, is that this strange arrangement manifests the degree to which whales, in New York in the early nineteenth century, could still be understood as exceptional, indeed, even as "wonders": they defied familiar categories in natural history, and were perhaps best treated as remarkable singularities, rather than as exemplifying any particular natural kind. As Sir John Falstaff said of the otter—invoking the clerical taxonomies that governed Friday meals—there were some creatures that were simply problematic: "she's neither fish nor flesh; a man knows not where to have her."[55] Vast and anomalous, whales figured significantly as creatures that defied easy identification, and they eluded the clear and distinct grasp of the inquiring eye and mind: not for nothing had the expression "very like a whale" come to mean in familiar English, "I don't know what you are talking about," or, more pointedly, "I don't believe you."[56] By these lights, it was perhaps best to leave the great whales outside of natural history classification altogether.

If a close examination of Scudder's Museum tells us anything about the status of whales in early nineteenth-century New York, it tells us that they were deeply unfamiliar creatures. This overarching fact presented the

the Californian Grey Whale in Icelandic Waters," *Investigations on Cetacea* 2 [1970], pp. 13–20). For these reasons (as well as the fanfare of 1814, which makes it clear that a whale jaw coming into the city made a splash) it seems very unlikely that Scudder had secured an additional whale jaw from a gray whale. Moreover, he would also have had to have sold the whale jaw he acquired in 1814, since no other jaw is listed in the 1819 inventory, but there is little reason to think he would have done this at any point through these years of expansion. Nevertheless, since "scrag" generally came to imply a whale having commercially valueless baleen, it seems strange that the jaw of a large mysticete whale could be called that of a "scrag," in view of the fact that rights and bowheads had the longest and most commercially valuable "whalebone" of all. A stumper.

[55] The line is from *Henry IV, Part I,* act 3, scene 3, where it functions as an off-color comment about the tavern hostess.

[56] "Very like a whale" is Polonius's whatever-you-say reply to Hamlet's musing on the shape of clouds in *Hamlet,* act 3, scene 2. The expression was a commonplace in the early nineteenth century. See, for instance, *Carolina Sentinel,* 20 March 1819, p. 4 (where the phrase serves as an italicized snicker after the report of a tall tale).

lawyers in *Maurice v. Judd* with a serious problem: the subject of this trial could not but be, in a very real way, a cipher. Anthon, pressing for the self-evident fish-ness of the whale, acknowledged that few in the room had ever had any direct experience of the creatures at issue, but he insisted that this must not lead the jurors to disqualify themselves as agents in the decision: "Many of us may not have seen a whale," he conceded, but this was no reason to be "led astray by the learning of philosophers." After all, the naturalists were willing to group the great whales with a set of other "non-fish" (as they would have it) that were much more familiar. Subjecting the naturalist's category of the "cetes" to a clever *reductio,* Anthon attacked:

> it is worthy of remark, that if the whale, by reason of his peculiarities, is to be removed from the finny tribe, the porpoise puts in an equal claim to this distinguished honor. . . . but the porpoise is an inhabitant of our own waters, we can judge his claim for ourselves; and as the porpoise is as much and no more a fish than a whale, in the acceptation of naturalists, we shall all have the demonstration of our own senses to keep us right on this great question.[57]

The plaintiff's side in the case was thus willing to concede the "peculiarities" of the naturalists—to wit, that the whale "breathes the vital air through lungs, that he has warm blood, that the whale copulates *more humano,* that the female brings forth her young alive and nourishes them at her breasts," and that "the tail is flat"—but they skillfully called into question, repeatedly and effectively during the course of the trial, whether dividing nature by such characters produced meaningful and reliable categories, categories that ought to displace the more familiar grouping according to the three "empires" of nature: "we stand forth," protested Anthon to the jury, "the advocates of the ancient empire of the whale, which, although fearfully shaken by the efforts of the naturalists, we trust will be established by your verdict."[58]

It is important to emphasize that by invoking this tripartite taxonomy and girding it with scriptural authority, Anthon and Sampson were not resurrecting an atavistic position (as we will see, Mitchill himself elected to buttress the work of Cuvier and Lamarck with his own gloss on Genesis), or pedaling a know-nothing natural history. Again and again during the course of the trial close knowledge of the natural world

[57] *IWF,* p. 13.
[58] Ibid., p. 14.

was deployed as the *solvent* of the taxonomic boundaries erected by the "new philosophy." For instance, the flatness of the tail, it could be shown, "was not quite peculiar to the whale, so as to distinguish it from the fish," being shared not only by the familiarly fish-like porpoise, but also by the flounder, a fish by any reckoning. This bit of quibbling could be answered, of course, by means of a lengthy explanation—namely by arguing "that the tail was not transverse with respect to the body, but would be perpendicular like that of other fish, if the fish swam perpendicularly, as others did; but as these fish swam on their side, the tail was in the same position: if they were to swim on their edge, then the tail would appear perpendicular."[59] But such exchanges revealed that fine work was necessary to maintain these novel classifications, and opened the way to mocking asides like Sampson's: "Is it known, sir, what the motive of that family is for following so odd a humor? It is, perhaps, out of pride to look like whales . . . ?"[60] Along the way the "new philosophy" began to smell very much of the lamp. A lamp burning very fishy oil.

Other differentia of formal taxonomy could wilt when subjected to the testimony of ordinary folk. Did bearing live young constitute a categorical distinction? In a showy gesture that surely played well in the gallery of the Mayor's Court, Sampson and Anthon plucked from the crowd a man dressed in a pea jacket who had been standing with "a number of persons in the garb of fishermen, whom curiosity had drawn into the court, to hear the subject of their occupation discussed by lawyers and doctors." Swearing him as a witness on the spot (was he really a well-coached player in a bit of courtroom theater? this is quite possible), the plaintiff's lawyers elicited from this seaman that it was a "fact within his own knowledge that the dog fish [meaning one of the various species of sand shark], brought forth its young alive; and that he had seen it pup in a boat." And yet its tail was fish-like, and it was never known to come to the surface to breathe.[61] Sampson, a master trial lawyer who had done his ichthyological homework, was able to adduce further that both eels and blennies were viviparous, and the former, like the whale, were scaleless.[62] Apparently, then, detailed natural history revealed that several of the ostensibly distinguishing characteristics of the cetes (according to the "new philosophy") did not, in fact, cleanly distinguish them from the fish.

These sallies at the categories of the naturalists left only delicate

[59] Ibid., p. 22.
[60] Ibid., p. 29.
[61] Ibid., p. 39.
[62] Ibid., p. 75.

matters as the rationale for keeping whales out of the fish category: breasts and reproductive organs. When called to the stand, Dr. Mitchill would make much of these points, which were at the heart of the classification he had come to defend: male fish, he could show, had "no *penis intrans*" and, unlike the cetes, propagation among the fishes involved the casting of "the fecundating fluid" over ova deposited in a "nidus."[63] These bedroom details, raised to the level of "philosophy," and set against the commonplace distinctions of the fish-market (and the sacred groupings of scripture), clearly struck many New Yorkers as grounds for a snicker: shortly after the trial, a satirical poem—there were to be a number of them—in the *Evening Post* invoked the city's preeminent naturalist with a sly dig at his prurient engagement with taxonomy, hailing him as

   Mitchell, who sung the amours of fishes . . .[64]

This chapter has argued that most New Yorkers in 1818 thought of whales as fish, indeed, even as the kings of the "empire" of the sea, the Genesical estate of the fishes. Contemporary comments made concerning the trial itself certainly suggest that this was the most common position, but further evidence can be found in the school texts of the period, and in the implicit taxonomies of the public institutions that installed an appreciation for natural history in the city. While there was understood to be much that was strange and anomalous about these creatures, perhaps wondrous, conceivably even monstrous and outside the bounds of ordinary categories, and while they were remote from the experience of most citizens, all these features only made the idea that whales belonged in the same taxonomic category with *human beings* even harder to credit. Whales seemed to be fish, though they were admittedly odd; that they were kin to man proved very hard to swallow. And yet this, it would emerge, was the claim of "those who philosophize."

---

   [63] Ibid., p. 29.
   [64] Drake, *Poems*, p. 28. The misspelling of Mitchill's name and the unusual conjugation of the verb "sing" are given here as they appear in the original.

THREE

The Philosophical Whale

SAMUEL LATHAM MITCHILL AND
NATURAL HISTORY IN NEW YORK CITY

Samuel Latham Mitchill, who had published an American edition of Erasmus Darwin's earthy *Zoonomia*, was not a man afraid to speak out loud about the loves of plants and animals; indeed, he was not a man afraid to speak out loud on most any topic. Almost thirty years after the doctor's death, in an eloge entitled *Reminiscences of Samuel Latham Mitchill, M.D., LL.D.,* Mitchill's younger contemporary, the memoirist and man-about-town John W. Francis, would recall that "It was a common remark among our citizens—'Tap the Doctor at any time, and he will flow.'"[1] If admiring, the comment captures something of a collective civic wink at the irrepressible energies, the polyglot enthusiasms, the distinguished eccentricities of New York's most publicly universal gentleman of the early nineteenth century, a man known variously as the "living encyclopedia," as a "stalking library," and (to his admired Jefferson) as the "Congressional Dictionary."[2] Because it was Mitchill above all who embodied the contentious new zoology in the trial of *Maurice v. Judd,* it is necessary to take a moment to review his biography, and to consider what he knew about whales and their place in nature.[3]

[1] Francis, *Reminiscences of Samuel Latham Mitchill,* p. 22.

[2] See Alan David Aberbach, *In Search of an American Identity: Samuel Latham Mitchill, Jeffersonian Nationalist* (New York: Peter Lang, 1988), p. 5.

[3] Mitchill has never been the subject of a full biographical study; insurmountable source problems probably preclude such a work (a fire destroyed the archive Mitchill himself had amassed to make such a book possible). My discussion of his life and work here has been informed by the manuscript collection of Mitchill's letters from Washington in the period of his service in the House of Representatives (held in the Manuscript Division of the Library of Congress), in addition to Mitchill material in the New-York Historical Society, the New York Academy of Sciences, and the Columbiana Collection (held in the Columbia University Rare Books and Manuscript Library). There are also several Mitchill letters in the DeWitt Clinton Papers (also at Columbia University Rare Books and Manuscript Library) which I have consulted. Period biographical material, besides the several works of John W. Francis, includes: Mitchill's own curriculum vitae, published as a pamphlet, *Some of the Memorable Events and Occurrences in the Life of Samuel L. Mitchill of New-York, from the Year 1786 to 1826* (New York: n.p., 1828); and Felix Pascalis's eloge, published as *Eulogy on the Life and Character of the Hon. Samuel Latham Mitchill, M.D.* (New York: American Argus Press, 1831). Note that the New York Pub-

For all his cosmopolitan aplomb, Mitchill was native to New York, born in 1764 as the son of a respectable Quaker farming family in the county of Queens, on Long Island. His medical education began with a two-year apprenticeship (1781–1783) to Dr. Samuel Bard, physician in the city of New York, who had himself studied at Edinburgh and encouraged his pupil to do likewise. The generosity of an uncle bankrolled Mitchill's three years overseas, and he returned to New York in 1786 with a medical degree from the "Athens of Britain," and the polish acquired during a tour on the Continent and a residency in London. He settled into the practice of medicine in the city, but soon found his attentions drawn to consuming questions of political order then the central preoccupation of the young Republic. Taking up the study of law, Mitchill came to the attention of Governor George Clinton, and entered public service as a commissioner in cession negotiations with the Iroquois Indians in the western part of the state. By 1790 he had been elected a state assemblyman, and his participation in a host of clubs and civic associations in the city over the next decade (in combination with the emerging friendship of DeWitt Clinton, a growing commitment to the Jeffersonian Republicans, and marriage to the prominent and wealthy Catherine Cock, widow of the successful shipbuilder Samuel Akerly), positioned this young professor of natural history at Columbia College (who had maintained a successful practice as a physician) to make a run for national office. In 1800 he was elected to the U.S. House of Representatives; four years later he would be nominated by the New York legislature to take over a vacated Senate seat, which he would occupy for five years. Additional terms in both the state legislature and the U.S. House would follow, bringing Mitchill back to New York City for good in 1813. From that point until his death in 1831 he was among the most prominent figures in the city: a founding member, member, honorary member, or corresponding member of some four dozen societies (a fact that did not go unmocked); a durable supporter of Mayor/Governor Clinton (whose controversial Erie Canal project he strongly endorsed); variously a professor in chemistry, botany, agriculture, and natural history; a projector in surveying and resource assessment in and beyond the

lic Library copy of the former, AN PV 78, contains a tipped-in supplement sheet that appears to be contemporary. Several secondary sources have been useful, including: Edgar Fahs Smith, *Samuel Latham Mitchill: A Father in American Chemistry* (New York: Columbia University Press, 1922); Hall, *A Scientist in the Early Republic;* and the recent work by Alan David Aberbach (which is mostly concerned with Mitchill's political career and vision for the Republic, and which does little with his scientific activities), *In Search of an American Identity.*

state of New York; a perennial voice on matters of hygiene, physic, and philosophies natural and moral, classically recondite and fashionably French. He founded journals, published his correspondence with men of learning in the United States and abroad, and endorsed a clutch of products and schemes that fit with his vision of health, national improvement, and ambitiously Jeffersonian ideals for the Republic. The most colorful effort to encompass the scope and verve of this multifarious philosopher may be a paragraph from Francis's *Old New York:*

> Ancient and modern languages were unlocked to him, and a wide range in physical science, the pabulum of his intellectual repast. An essay on composts, a tractate on the deaf and dumb, verses to Septon or to the Indian tribes, might be eliminated from his mental alembic within the compass of a few hours. He was now engaged with the anatomy of the egg, and now deciphering a Babylonian brick; now involved in the nature of meteoric stones, now on the different species of brassica; now on the evaporization of fresh water, now on that of salt; now offering suggestions to Garnet, of New Jersey, the correspondent of Mark Akenside, on the angle of the windmill, and now concurring with Michaux on the beauty of the black walnut as ornamental for parlor furniture. In the morning he might be found composing songs for the nursery, at noon dietetically experimenting and writing on fishes, or unfolding a new theory of terrine formations, and at evening addressing his fair readers on the healthy influences of the alkalis, and the depurative virtues of whitewashing.[4]

In short, he aspired to be a Masonic sage of the Scottish Enlightenment, one who deployed his wisdom in the service of a patriotic Jeffersonian nationalism. His misfortune, historically speaking, was finally to achieve this lofty goal in the very same years that saw both these projects become quaint relics of another era.

It is this early nineteenth-century watershed that accounts in large part for the darkening tincture of ostentatious respect and patronizing bemusement that attended Mitchill in his final decade (and a hint of which can be discerned in the trial of *Maurice v. Judd*): the founder of institutions became himself an institution; the collector of artifacts became, in a manner of speaking, an artifact. By the end he could be eulo-

[4] Francis, *Old New York,* pp. 90–91.

gized as "one who knew all things on earth and in the waters of the deep," even as it was acknowledged that the breadth of his erudition (and his taste for pomp and posture) made him, truth be told, more effective as a promoter of the sciences than as a proper contributor to that enterprise, an enterprise which was increasingly understood—in the terms dictated by Benjamin Silliman and Joseph Henry—as the domain of sober specialists, rather than flamboyant gentlemen-*érudits*.[5]

It might be argued that 1818–1819 represented the very cusp of this change, and that the trial itself (as I will argue in the conclusion) played a role in ushering Mitchill from his ascendancy; but what is essential to recall in looking at Mitchill's testimony in *Maurice v. Judd* is that in December of 1818 Mitchill was, without question, the most prominent man of science in the state of New York, an unparalleled authority on questions of natural order and a powerfully positioned patron-administrator of the institutions of learning on the fair island of Manahata. Moreover, he would remain close to the vanguard of the scientific culture of the city for some years to come: unlike other philanthropic naturalist-improvers of his age (for instance, DeWitt Clinton himself, who in many ways gave shape to the ideal in New York in this period) Mitchill retained a close connection to an emerging younger generation—men like Peter Cooper, John Torrey, and James Ellsworth DeKay—whose idea of a scientific society left no room for coin collecting or second-rate philology. Mitchill's Lyceum of Natural History was the gravitational center for the "zeal of these aspiring youths" (to quote John Pintard, who warned presciently that this clique would "eclipse the old Societies"), and the efforts of these upstarts to keep Clinton out of their new society has rightly been seen both as a symbolic changing of the guard, and as an indication that Mitchill—who objected to the exclusion of his longtime friend and ally, but who chaired the meetings that voted the blackball—had the mettle to endure, despite his patina, in a way that distinguished him from most of the other men of his generation.

The explanation for his good standing in this "proto-professional" scientific community lies in the great success of Mitchill's lectures in

---

[5] The quote is from Hall, *A Scientist in the Early Republic*, p. 18. On Silliman, see Chandos Michael Brown, *Benjamin Silliman: A Life in the Young Republic* (Princeton: Princeton University Press, 1989). On Henry, see Albert E. Moyer, *Joseph Henry: The Rise of an American Scientist* (Washington, DC: Smithsonian Institution, 1997); and Thomas Coulson, *Joseph Henry: His Life and Work* (Princeton: Princeton University Press, 1950). George H. Daniels discusses these figures, and the more general forces at work in the transition, in: "The Process of Professionalization in American Science: The Emergent Period, 1820–1860," *Isis* 58, no. 2 (1967), pp. 150–166.

natural history at the College of Physicians and Surgeons. It is no exaggeration to say that Mitchill had himself built New York's nascent community of scientific specialists through his teaching. Of the twenty-one founding members of the new Lyceum, for instance, twelve were associated with the new college, and at least ten of the early members were either students or recent graduates—all of whom had attended Mitchill's celebrated, synthetic, hands-on, and even "research-oriented" course, which met daily from one to two p.m., beginning the first Monday in November and running through to the last day of February.[6] This amounted to more than ninety "discourses," presented to classes ranging well over fifty students, covering minerology, botany, and zoology, and demanding field trips to Hosack's botanical garden in the northern suburbs (now Rockefeller Center), as well as visits to several natural history cabinets in the city, including, as we have seen, Scudder's. The course had a long life span. Begun in 1809 (when Mitchill was briefly back in New York between legislative appointments), it ran with few interruptions to the end of Mitchill's active life: in 1826, at the age of 62, he wrote with pride to Clinton that "I have been successful in the course of lectures to an unprecedented degree," and that, with eighty presentations behind him, he had yet to miss a single meeting.[7]

It is a testimony to the reputation and impact of this course that several sets of student notes taken by Mitchill's pupils survive, not to mention at least one not-so-flattering notebook-doodle caricature of the great (and bulbous) man in action on the podium [Plate 5]. These materials (admittedly more the former) offer a superb opportunity to understand how Mitchill configured the natural world, and how he taught the classificatory sciences that he brought to the bar in *Maurice v. Judd*. The notes reveal that Mitchill presented lectures dynamically engaged with contemporary continental authorities and changing ideas about the earth and its inhabitants, lectures that propounded the virtues of the new comparative anatomy while underscoring the significance of natural history for a young nation. It was a course designed to attract and prepare natural history practitioners, and to ensure that they were familiar with the most up-to-date work being done in the field.

---

[6] See the announcement in *American Monthly Magazine and Critical Review* 1 (1817), p. 121. The term "proto-professional" is used by Thomas Bender in reference to this group. "Research-oriented" is admittedly somewhat anachronistic, but defensible, I think, since Mitchill explicitly hoped that many of his students would eventually contribute to the production of knowledge about the country.

[7] Mitchill to Clinton, 20 Feburary 1826, DeWitt Clinton Papers, Columbia University Rare Books and Manuscript Library.

For instance, G. S. Townsend, who attended the 1816–1817 course, carefully inscribed in his notebook at the start of the lectures on the animal kingdom: "The Parisians have gone far ahead of any other people in zoology," and beside that Townsend copied down the mantra of the new taxonomy, "Zoology is built upon the base of anatomy."[8] He went on to record some of Mitchill's statements verbatim, or at least in the professor's own voice, writing: "In tracing an[s] [animals] I shall begin, with nature, with the most simple. I have the authority of Lamarque, Dumeril + Cuvier."

Such references to "authority" were not at all deployed to foreclose debate. On the contrary, Mitchill embarked on the taxonomy of the animal world by underlining that even the basic boundaries of this seemingly self-evident grouping were by no means uncontested. Reflecting a familiarity with Lamarck's work on the "animaux sans vertèbres," Mitchill presented the seemingly fundamental category of "animals" as itself a fundamental problem. Quoting again from Townsend's notes:

What is an animal? It may appear plain to distinguish . . . the horse from that on which he feeds. but so mysterious are the incipient traces of animality that it is one of the most puzzling questions to distinguish anl. [animal] from veg[etable]. The stom[ach] is no critereon for there are whole classes without stom[ach]. or a gastric! An'l [animals] are said to have a nervous system, but there are large classes without brain or nerves! Generative orgs. [organs] were supposed characteristic, i.e. organs whereby an an. [animal] was enabled to perpetuate their own species, but there are anl' [animals] without them & which propagate their kind like vegetables i.e. by being split up! Neither is locomotion charact[c] [characteristic] for the oyster is fixed . . . [9]

The exclamation points prick young Master Townsend's ledger book throughout, and they fairly bark Mitchill's gift for imbuing his subject with urgency and surprise.

An account like this emphasized above all that taxonomy could be a matter for argument and debate, and that natural history was by no means a static collection of facts. Another of Mitchill's students, commenting

[8] Townsend Manuscript, third lecture, Columbiana Collection, Columbia University Rare Books and Manuscript Library. Compare the notes by Thomas A. Brayton (MS 70–1713) in the collection of the New York Public Library, Manuscripts Division.

[9] Townsend Manuscript, second lecture, Columbiana Collection, Columbia University Library Rare Books and Manuscript Library.

on an earlier version of the class, emphasized how Mitchill continuously and explicitly positioned his own views with respect to other authorities, and always made clear to his students how and why such departures were necessary. For instance, Mitchill espoused his own mineralogical terminology (closely related to his ideas about the significance of oxygen and phlogiston, which he treated extensively in his work on "septon" and the virtues of the alkalis), and he reconfigured the classes within the Linnaean botanical scheme (placing the "cryptogamia"—plants like the mosses that did not appear to grow from seeds—at the head of the order). "This innovating temper of the professor is not limited to systems of mineralogy and botany," wrote one pupil, commenting on Mitchill's style after taking the course, "he carries it with him into his zoology, and he transposes the classes of animals with still greater freedom."[10] This meant, as this former student went on to explain, that "he follows in the main the system of Cuvier, who founded the classification upon their internal organization and anatomical constitution; but he inverts and transposes it for the purpose of greater perspicuity."[11]

This "inversion" and various other transpositions all aimed to make the different branches of natural history conform more closely to a stepwise *scala naturae*, whereby Mitchill's lessons (mirroring the Divine plan) could proceed from inanimate matter, up through plants in order of increasing complexity, and then into the animal kingdom, beginning with the most plant-like creatures, and building up to the mammals, under which heading he discussed the orders in such a way as to "climax" with his lectures on human beings (order, *Primates*)—the topic with which the later versions of the course closed. Mitchill particularly prized his finale: a set of rousing discourses on the "aboriginals of Fredonia" and the natural history of mankind.[12]

What is significant about all this is less that Mitchill continued to use

---

[10] A Correspondent, "Cultivation of Natural History," p. 159.

[11] Ibid.

[12] It is evident that the course evolved over Mitchill's life, and that its structure changed: for instance, it appears that early on the class ended not with the section on man but with an additional sequence of lectures on "uranology," or astronomy. See "Outline of Professor Mitchill's Lectures on Natural History, in the College of New-York, delivered in 1809–1810, previous to his departure for Albany, to take his seat in the Legislature of the State," *The Medical Repository* 13 (1809–1810), pp. 257–267 (NB: This journal has a complicated publishing history; a digital edition is now available, but I have found that the ProQuest citation format is not reliable; I have followed the citation format of Hall, *A Scientist in the Early Republic*). It is interesting to consider the extent to which natural history in this period feels like a discipline organized by its pedagogical obligations. The "order of nature" turns out to conform with uncanny precision to the "order of a year-long lecture class."

a version of the (increasingly outmoded) "great chain of being" as the armature of his course, than that he insisted upon teaching natural history as a subject demanding judgment and argument, a subject in which dissent was possible, and new ideas and new facts mattered.[13]

Mitchill also knew how to endow the subject with urgency and theater. His object-centered teaching put materials before students' eyes and even in their hands.[14] For instance his opening lecture was built around three carefully chosen objects, tokens of the three kingdoms of nature, but each endowed with a particular power: for the animal kingdom, Mitchill would declare, "I present you with the weapon of the stingray," before dilating on the local species found in the Long Island Sound, the dangers occasioned by the animal's wound, and the fact that it was safe to eat, should his young listeners ever find themselves shipwrecked (or, for instance, in France, where, Mitchill the cosmopolite noted, it was "eaten voraciously"); the mineralogical specimen was a lump of arsenic ore from a large vein up the Hudson Valley, which contained "enough, as I told a French Chemist here, to poison the universe" (a fact calculated to seize the youthful imagination), and Mitchill went on to discuss the importance of arsenic in the molding of spherical shot (a proper topic for a lecturer who held the office of "Surgeon General of the Militia" in New York, and who had assisted in drafting the city's fortifications during the War of 1812); finally, as a specimen from the vegetable world Mitchill offered a sample of "spurred rye" or ergot, and he gave his opinion of its use in accelerating childbirth. Central to this showy opening lecture was the fact that each of these was a "sample of American productions," for Mitchill made much of the significance of natural history in the United States, where, as he put it, "beauty, novelty, and sublimity characterize creation to a much greater degree than in any part of the old world"—it was, he explained to his class, "a fortunate occurrence for a student of nature, to be born in the United States."[15]

[13] On this notion of continuity in the organic world, see the classic study (originally composed in the 1930s), Arthur O. Lovejoy, *The Great Chain of Being: The Study of the History of an Idea* (Cambridge, MA: Harvard University Press, 1966).

[14] For a clear statement of his pedagogical philosophy, see the pamphlet published in the 1820s by the College of Physicians to promote Mitchill's course, which closes as follows: "It is the invariable practice of the Professor to teach by specimen, picture, map, diagram, table, &c. to the whole extent of which such representations are susceptible; under a conviction that material objects aid most impressively the abstract conceptions of the understanding." College of Physicians and Surgeons in the City of New York, *A Concise Memorandum of Certain Articles Contained in the Museum of Samuel L. Mitchill* (New York: E. Conrad, circa 1825).

[15] A Correspondent, "Cultivation of Natural History," p. 161. Mitchill's slightly truculent enthusiasm here and elsewhere must always be understood as a tacit rejoinder to the thread of

A dynamic science, taught with objects, central to the life of the new Republic: Townsend's notes for zoology lecture 8 of 1816 wonderfully exemplify how Mitchill's course wove together these three preoccupations. Announcing to the students that an exciting discovery would have to preempt the ordinary lecture, Mitchill proceeded to read aloud a letter he had just received from a state surveyor at work in western New York, who had sent along three pickled fish-skins from the Finger and Great Lakes. Displaying one of the skins, Mitchill announced that it appeared to be in the family of the cod, though the fish had come from a body of fresh water in the interior, a fact that prompted deep reflection about terrestrial and organic change. Townsend copied Mitchill's grave remarks: "he was an inhabitant of the ocean when there and has survived the changes that have occurred and now lives in fresh water, but still lean and poor. He never grows yet though he propagates his kind." Here was an unexpected lesson: proof of profound secular change resided at the bottom of the western lakes. And Mitchill knew how to endow the mutability of the earth with political significance for the nation: he famously marshaled the idea of the historical contingency of geology as an argument in support of Jefferson's Louisiana Purchase, since it was in the nature of landforms to change, with or without the consent of nations.[16]

This close review of Mitchill's teaching shows how he built a celebrated course that dramatized natural history as a vigorous and important enterprise. By doing so he became the most public scientific figure in the city of New York in the early nineteenth century, and drew around himself a community of younger naturalists taken with the investigation of the natural world. But the notes scribbled by Mitchill's students do more than lift the curtain on his podium performances. They also demonstrate the degree to which he continued to develop his thinking long after his own student years in Europe. The work of Lisa Rosner and others on medical education in Edinburgh in the late eighteenth century conveys a clear sense of the kind of training that Mitchill would have received in chemistry (where he studied with Joseph Black, who stimulated Mitchill's

argument in European natural history that asserted American plants and animals lacked size and vigor (on account of climate, etc.). For a résumé of this long-standing debate, see Antonello Gerbi, *The Dispute of the New World: The History of a Polemic, 1750–1900* (Pittsburgh: University of Pittsburgh Press, 1973).

[16] "[I]f by any force of the currents of the ocean, or any conflicts of the wind and the waves, a new surface of earth should emerge from the neighborhood of Cape Hatteras, would it be unconstitutional to take possession of it?" Cited in Hall, *A Scientist in the Early Republic*, p. 114.

interest in the phlogiston debates), in botany (taught by John Hope, who was at the end of his career during Mitchill's tenure), and most importantly, in anatomy, which was taught in those years by Alexander Munro "secundus" (as he was known), the most popular of the great dynasty of Edinburgh "philosophical anatomists." That course ended with a survey of comparative anatomy, and it is safe to assume that it was in those lectures that Mitchill had his first rigorous exposure to the anatomical basis of animal classification (Parisian natural history was at this time represented better in Edinburgh than anywhere else in the English-speaking world).[17] But while Mitchill would later write in praise of his Edinburgh education, invoking his "able masters" and musing on his "tours around that great seat of learning" and his "excursions to the mountains" to study geognosy, he was acutely aware of how rapidly things had changed in natural history in the years since he had returned. Recalling his student days, Mitchill wrote in 1818 that

> During several visits to London I became an industrious visitor to museums, libraries, galleries, and even the environs of the city. The rapid and increasing march of knowledge since I was there has outdone all former example.

And the same could be said for Paris:

> While I resided in Paris, I endeavored to acquire as much information as possible from the admirable institutions there. But the present constellation of science had not then risen.[18]

These comments highlight the degree to which Mitchill's engagement with continental developments in natural history was a product of his *reading*. Books kept him abreast of the new French comparative anatomy, and it was out of his store of fresh-from-the-press texts that he built his lectures; the same books served as the references at his elbow as he composed his monographic studies in natural history, saliently his

[17] Lisa Rosner, *Medical Education in the Age of Improvement: Edinburgh Students and Apprentices, 1760–1826* (Edinburgh: Edinburgh University Press, 1991). See particularly chapter 3 for a discussion of the curriculum in the relevant years, and pp. 47–48 for Munro's anatomy class. Also relevant, for context on the American graduates from Edinburgh who returned to the United States to practice, is Lisa Rosner's "Thistle on the Delaware: Edinburgh Medical Education and Philadelphia Practice, 1800–1825," *Social History of Medicine* 5, no. 1 (1992), pp. 19–42.

[18] Georges Cuvier, *Essay on the Theory of the Earth,* edited by Robert Jameson, with additions and a new introduction by Samuel Latham Mitchill (New York: Kirk and Mercein, 1818). See Mitchill's introduction.

"Fishes of New-York." While no inventory of Mitchill's library survives, hints lie scattered in various places, making it possible to recover a number of Mitchill's key texts, and even to catch glimpses of how he used those books in the practice of natural history.

For instance, in 1818, shortly before the trial of *Maurice v. Judd,* Mitchill published an American edition of a work by Cuvier entitled *Essay on the Theory of the Earth;* this was in fact a reprint of the translation of Cuvier's "Preliminary Discourse" that had been commissioned by Robert Jameson and published at Edinburgh with his notes and commentary in 1813. To this already composite text Mitchill added an extensive appendix on the relevant North American fossils, and on the potential importance of New World specimens to Cuvier's ideas. While this might suggest that Mitchill's contact with French works was wholly mediated through (questionable) English translations, this was not the case. In fact, in 1807 Mitchill published a detailed abstract of the 344-page *Zoologie Analytique* of Constant Duméril, a work that had appeared less than a year earlier in Paris, and that engaged closely with the classifications of both Geoffroy and Cuvier. This indicates the speed with which such works could cross the Atlantic and, via Mitchill, become part of the conversation around natural history classification in the United States.[19] It is clear from other sources that Mitchill had read Cuvier's three-volume *Leçons d'Anatomie Comparée* of 1805 in addition to that author's older *Tableau Elémentaire de l'Histoire Naturelle des Animaux* (1797).[20] So attentive was Mitchill to the importance of staying on top of the classificatory enterprises of the Paris Museum, that no sooner had Cuvier's four-volume *Règne Animal* come off the presses than Mitchill preoccupied himself with securing a set for the library of the new Lyceum of Natural History. It was, in the end, the first book for which the society paid out of collective funds.[21] They accepted donations, of course, but

[19] It is interesting to review advertisements in the New York papers from this period for insight into the availability of continental natural history publications. For instance, see the posting in the *Commercial Advertiser* of 12 February 1819 by a bookseller with the improbable name of Fernangus de Gelone, who announced that he had for sale "A great quantity of the best and latest publications on medicine, mineralogy, botany, natural history, and mathematics" including Buffon, La Treille, all twelve volumes of Humboldt's *Personal Narrative* (this would seem to have been a French edition, since the English ones were all shorter, as far as I know), and an edition of Aristotle's "On Animals" in the original Greek.

[20] The latter was in the library of the cabinet of the Literary and Philosophical Society. See "Donations for the Library and Cabinet," *Transactions of the Literary and Philosophical Society of New-York* 1 (1815), in the backmatter.

[21] Minutes of the Lyceum of Natural History for 4 January 1819, collection of the New York Academy of Sciences.

such choice and important volumes could not reliably be secured merely by waiting for a fortuitous gift.

Nor did Mitchill simply file such texts on the shelf or merely crib from them in preparing his lectures. He *used* them energetically in investigating specimens, and in evaluating the productions of American nature, stacking volumes open on his desk and setting their assessments against each other. In doing so he pointed up their strengths and weaknesses, and composed works of his own which were in conversation with these books, and supplementary to them. The best glimpse of this active process is provided by the original work that cemented Mitchill's reputation as an innovator and contributor in the science of classification, "The Fishes of New-York," a 150-page illustrated monograph ichthyological published in 1815 in the first volume of the *Transactions of the Literary and Philosophical Society of New-York,* and updated and expanded over the next ten years in a series of supplements issued in various other journals. In the preface to the original work, Mitchill offered a valuable sketch of the process of collation and bibliographical scrutiny that supplemented his early morning expeditions to the fish market:

> I have had before me, during my inquiries, the Leyden copy of the *Museum Ichthyologicum* by L. T. Gronovius, 1754 fol., Castel's French edition of Bloch's *Histoire Naturelle des Poissons,* Paris, 10 tom. 12mo 1801; Gmelin's edition of Linné's *Systema Naturae,* with Turton's English translation; and the ichthyological part of Shaw's *General Zoology,* as published in London.[22]

Mitchill wrote that these, along with "Catesby and Edwards," were "constantly at my elbow." By evaluating the skeletal character (bony or cartilaginous) of his specimens, and counting the rays of their fins, Mitchill was able to assign each of his New York fish to one of the six orders recognized by Shaw, and to compare the products of the New York waters with the species and varieties enumerated by European authorities. It was said that Cuvier himself praised "The Fishes of New-York," and there is good evidence that Mitchill maintained a private correspondence with Lacépède in the years that followed its publication, Lacépède being the author of the standard five-volume *Histoire Naturelle des Poissons.*[23]

---

[22] Samuel Latham Mitchill, "The Fishes of New-York," *Transactions of the Literary and Philosophical Society of New-York* 1 (1815), pp. 355–492, at p. 358.

[23] Since this work was published between 1798 and 1802, it is noteworthy that Mitchill omits mention of it in his list of sources. The almost reliable confidence-man-*cum*-naturalist Constantine S. Rafinesque ("Introduction to the Ichthyology of the United States," *American Monthly*

That Mitchill appears not to have had Lacépède's texts "at his elbow" as he worked up his fish specimens suggests that, for all his efforts to present the Fredonian naturalist as entirely engaged with the texts, debates, and personalities of the Anglo-European scientific community, there were real difficulties practicing natural history at such a remove from the libraries and cabinets of London and Paris. Mitchill was by no means ignorant of these disadvantages, but he advocated a program of collective research on the natural history of the Republic (centered on New York) that would take advantage of the compensatory benefits of the location, social structure, and mercantile energies of his beloved city and its surroundings. Here Mitchill and DeWitt Clinton shared a vision for the sciences in New York and the United States. As Clinton argued concerning New York in his "Introductory Discourse" of 1814:

> We may say of this place as Sprat, in his History of the Royal Society, said of London: "it has a large intercourse with all the earth; it is, as the poets describe their house of fame, a city where all the noises and business in the world do meet and therefore this honor is justly due to it, to be the constant place of residence for that knowledge which is made up of the reports and intelligence of all countries."[24]

And out of the spark and whirl of continuous, competitive merchant reconnaissance, enlightenment would blaze, since "science, like fire, is put in motion by collision."[25]

Mitchill placed this ideal—of a science kindled by the friction of commerce, and ventilated by the commonweal—at the center of his program for an American natural history. It was this vision, spangled with a democratic and populist enthusiasm, that came to the fore in Mitchill's rhetoric on the occasion of the founding of a short-lived sci-

---

Magazine and Critical Review 2 [1818], pp. 202–207, at p. 202) would comment in 1818 on the inadequacy of Lacépède for New World species, but that leaves open whether Mitchill had the book to hand as he worked up his specimens in the early teens. For a brief (and not wholly satisfactory) discussion of Mitchill's continental correspondents and readers, see Hall, *A Scientist in the Early Republic*, pp. 85–86. Work in French archives could presumably sort out these issues. For an interesting discussion of Lacépède's fish taxonomy, see Alan H. Bornbusch, "Lacépède and Cuvier: A Comparative Case Study of Goals and Methods in Late Eighteenth- and Early Nineteenth-Century Fish Classification," *Journal of the History of Biology* 22, no. 1 (Spring 1989), pp. 141–161.
[24] DeWitt Clinton, "An Introductory Discourse," *Transactions of the Literary and Philosophical Society of New-York* 1 (1815), pp. 19–184, at p. 47.
[25] Ibid., p. 49.

entific society at the turn of the century, when he announced that the aim of the organization was to "arm every hand with a [geologist's] hammer, and every eye with a telescope."[26] Where zoology was concerned, Mitchill spelled out a similarly inclusive program in an interesting circular, printed by the New-York Historical Society in 1817. The purpose of this flyer was to drum up support for the Society's natural history collection, and to expand a network of informants and contributors around New York and beyond: In the pursuit of "facts, specimens, drawings and books" Mitchill put out the call "that all citizens may be solicited to exert themselves . . . that much may be accomplished with very little cost." Because the city of New York "may be considered as a center surrounded by wonders" of minerals, fossils, plants, and animals, "all hands must be employed" in the harvest. Mitchill then went through the different classes of zoology, addressing himself, taxon by taxon, to the relevant tradesmen and merchants: Where ichthyology was concerned, the manifest local and national importance of the New York fisheries meant that "no additional recommendation is necessary, further than to ask of our fellow-citizens all manner of communications." Continuing, he exhorted that "contributions toward the natural history of the *Mammalia* may be expected from the fur merchants, furriers, and hunters. Almost everything known under the titles of FURS and PELTRIES passes through our city or is contained within it." Moreover, he went on, "zoological research is promoted in several ways by foreign commerce. Living animals are frequently imported . . . and cargoes, and even ballast, often contain excellent specimens, both of the animal and the fossil kinds." Even the scrapings from the hulls of vessels ought to be perused "before they are disturbed for the purpose of cleaning and repairing." Nor was this collaborative project limited to passing along specimens to the central cabinets of the city. Mitchill wanted the citizens to become practitioners of the most up-to-date approaches in zoology. The circular explained that

> Anatomy is the basis of improved zoology. The classification of animals is founded upon their organization. This can be ascertained only by *dissection*. The use of the knife is recommended for the purpose of acquiring an acquaintance with the structure of animals. It is proposed that the members avail themselves of all opportunities to cultivate COMPARATIVE ANATOMY, and

[26] Hall, *A Scientist in the Early Republic*, p. 71.

to communicate the result of their labors and researches to the society.[27]

At his death in 1831 Samuel Latham Mitchill could be remembered as having been the patriarch of natural history in New York City for more than thirty years. This chapter has thus far reviewed the ideas, activities, and institutions that he nurtured during the first decades of the nineteenth century in order to show several things: first, that Mitchill's teaching and publishing aimed to build a community of natural history practitioners in the United States in this period, practitioners who were informed about the latest teachings of the new zoology coming out of the Paris school, and who could contribute to the global project of taxonomy and comparative anatomy; second, that Mitchill conceived of a collaborative and distributed program of natural historical investigations in the United States—one centered in a particular way on New York City—that would not only take advantage of the proximity of a distinctive American nature, but would also tap the mercantile circulations that flowed through the commercial heart of Manhattan. The citizens of this community, mustered to this collective project, would reap material, intellectual, and even spiritual benefits from the enterprise. Indeed, specifically moral edification was not to be overlooked. As one of Mitchill's students put it: "There was a wholesome natural theology, blended somewhat after the manner of Paley," in Mitchill's lectures.[28] Mitchill himself, speaking of a proper cabinet in natural history (where the whole of the terraqueous globe might be drawn together), averred: "Let the philosopher survey the whole and draw wise and pious conclusions."[29] And not just philosophers: as the New-York Historical Society's circular of 1817 demonstrates, so deeply was Mitchill committed to teaching the Republic about the virtues of comparative anatomy, that even before the case of *Maurice v. Judd*, he had taken to postering the city with broadsheets encouraging the citizenry at large to take up the scalpel in the pursuit of the order of nature.

[27] Circular issued from the New-York Institution: Samuel L. Mitchill, "American Zoology and Geology," dated March of 1817. This may exist in other collections, but the only copy I have seen is the one tipped in to the Minute Book of the New-York Historical Society, circa 1818 (manuscript collection of the New-York Historical Society).

[28] Francis, *Reminiscences of Samuel Latham Mitchill*, p. 16.

[29] Mitchill, "American Zoology and Geology." For the provenance of this document, see footnote 27 *supra*.

This, then, was the person—and the program—that took the stand in the Mayor's Court a little after five p.m. on the 30th of December, 1818, as the star witness in the case of *Maurice v. Judd,* a case that placed a question of classification and comparative anatomy in the primordial democratic arena, a jury trial; a case that, in the process, put Mitchill's vision for natural history in the United States on the stand as well.

This "ornament to his country" (as opposing counsel John Anthon called him) was late. When he arrived, he apologized for keeping the court waiting, and without further delay he embarked upon a defense of Judd's claim that a whale was not a fish, grounding that claim, from the very start, in the kind of democratic and centripetal natural knowledge that Mitchill passionately espoused. He announced:

> New-York is a point into which much information centers. Men departing from this point circumnavigate the globe, voyaging from the arctic to the antarctic regions. From this class of my fellow citizens, much of the information I possess on this subject has been derived; and as a man of science, I can say positively, that a whale is no more a fish than a man; nobody pretends to the contrary now-a-days, but lawyers and politicians.[30]

It was an opening statement calculated to emphasize not his Edinburgh erudition, or his continental correspondents, but his cherished image of natural-historical New York, poised at the center of a network of worldly informants. At the same time it was, to an ambitious degree, an assertion that aimed to ally his position not with the learned elites of the city, but rather with the world of the docks and the Tontine Coffeehouse (the meeting place at Wall and Water streets where speculators and magnates rubbed shoulders and cut shipping deals), the world of the merchant traders, seal-hunters, and tars, with their global reach.[31]

But it was also a stance that opened Mitchill to a savvy cross-examination, one that undermined his effort to ventriloquize the "everymen" of maritime New York, and that began instead to maneuver him onto the more circumscribed authority of the suspect and alien "new philosophy" of Parisian zoology. Mitchill's primary antagonist, William Sampson (who, like his co-counsel Anthon, adorned his examination with ostentatious reverence for Mitchill's sagacity), pressed the doctor on

[30] *IWF,* p. 26.

[31] On this community see: McKay, *South Street;* and Robert Albion, *The Rise of New York Port, 1815–1860* (New York: Scribner's Sons, 1939).

the question of just what, exactly, "common" wisdom instructed on the question, forcing Mitchill to acknowledge, with some exasperation, that

> The great bulk of mankind that speak English, would call a whale a fish, and they would say the same of a crab or a clam, and with them I would not dispute the question. If I was to go into the market amongst my Long-Island friends, I would not debate the question whether the lobster were a fish or a crustaceous animal, or whether a clam were a shell fish or a mollusca.

And thus, Mitchill conceded, it might happen that laws would be framed in ways at odds with the lineaments of nature, since, as he noted proudly

> The legislature, to the honor of our democracy, consists of all classes of men. It is one of the felicities of our form of government, that all classes are represented.[32]

From the very start of his testimony, Mitchill tried to stand his position on a vernacular natural history embedded in his vision of a dynamic, worldly, democratic republic where the knowledge of nature was integral to practical progress and national virtue. As this opening exchange suggests, it would be a tricky matter. Countering this populist position, it would be the art of the plaintiff's lawyers to depict Mitchill's taxonomy, by contrast, as a foreign import, an unseemly innovation, and as the peculiar domain of an eccentric minority of "philosophers." At stake, then, was more than merely the question of whether whales were fish. At stake was the status and standing of natural history—the value and place of scientific knowledge—in the Republic.

Mitchill's second effort to buttress his "paradox" drew in a similar way on the familiar, rather than on invocations of exogenous or specialized learning: from the world of maritime New York, Mitchill moved to the tradition of pious natural theology within which he always sought to work. Answering Anthon's opening argument about the taxonomy of Genesis, Mitchill asked the court to return to verses 20 and 21 of chapter 1. Taking the Bible in hand, Mitchill read the verses and offered his own gloss on the lines:

> And God said, Let the waters bring forth abundantly the moving creature that hath life, and let fowl fly above the earth in the open firmament of heaven.

[32] *IWF*, p. 26.

> And God created the great whales, and every living creature that
> moveth, which the waters brought forth abundantly . . .

From these two lines Mitchill argued that "the formation of whales was
a distinct exertion of the creative power from the creation of fishes,"
since the waters "brought forth abundantly" *before* the specific invocation
of the creation of whales. For Mitchill, that initial "teeming" represented
the creation of ordinary fish, and the whales were mentioned separately
in the next verse because of their distinctive nature.

This Talmudic exegesis met with skepticism from Sampson: after all,
he protested, the word "fish" nowhere appeared in the initial creative act.
To which Mitchill replied: "The word fish may not be used, but the inference is obvious." At this point the recorder himself, the judge, intervened,
for clarification, asking Mitchill if he meant to assert that "on the fifth day
these large cetaceous animals were formed by a distinct exertion of Divine
power, and were of a distinct formation?" To which Mitchill replied:

> I do. It was a distinct creation by the Almighty power after he
> had created the other marine animals, although all were made on
> the same day. I mean to say, that from this it may be implied, that
> the cete was a distinct creature, and so the able writer of that
> chapter has described it; and a great man he was. He knew the
> difference between a whale and a fish. It is a luminous text, and
> displays great learning, and throws great light upon this subject.[33]

Sampson suggested that there would be occasion to return to scriptural
classifications before the trial ended, and the questioning turned to the
technical matters of classification, and the scientific basis for Mitchill's
shocking opening assertion, "that a whale is no more a fish than a man."

"No More a Fish than a Man"

Before turning to the heart of Mitchill's taxonomic testimony, and
Sampson's efforts to undermine it, we need to take a moment to consider
just what Mitchill himself knew about whales. What was the status of
the cetes in the classificatory texts that Mitchill knew, and what additional access did he have to these unusual animals?

We are fortunate to have an invaluable source as a point of departure
on this question: Mitchill's own lecture notes for that portion of his
course in natural history, notes that look to have been drafted in the

[33] Ibid., p. 28.

1810s and supplemented with later annotations.[34] While Mitchill's three surviving lecture notebooks are not complete, they do include the relevant section of the "Mammalia" lectures, in which Mitchill emphasized the distinguishing characteristics of the class as a whole—"they give suck and are viviparous; all the ova they have if any are in the ovaria, the evolution of it [*sic*] depends upon the stimulus of a fluid emitted from an instrument of the male called penis intrans, have double hearts, a warm red blood circulating through aortic, pulmonic and hepatic systems"—before turning to the first order, the *Cetae*. There, Mitchill opened his lecture to his students with a discussion of the very problem that would be before the court in 1818:

> It may seem strange to you that whales are not put under fish, for that is the common term applied to them—but they are not fish, they breathe thro' spiracles by lungs, copulate by penis intrans, have warm red blood <[e] teats near the anus > & suckle their young by teats. Character: spiracles for breathing placed on forepart of scull [*sic*]—come up occasionally to breathe by these, have no feet, pectoral fins without nails or claws, tail horizontal, fish all have perpend. tails.[35]

He went on to list a dozen species, including the narwhale, the sperm, the *mysticetus,* and the smaller dolphins and porpoises.

This source gives us a very clear sense of the position Mitchill gave to the cetaceans in his teaching at the College of Physicians and Surgeons, and the confidence with which he instructed students that whales were not fish. Where did Mitchill get this idea?

For starters, it is true that, by 1818, the preponderance of zoological authorities had ceased to classify whales and dolphins with the fish. But the process of that move had been slow, it remained incomplete, and, as I suggested in chapter 1, the move itself had been central to the reformulation of zoological classification in the second half of the eighteenth century. The seventeenth-century English anatomist John Ray—whose joint work with Francis Willughby, *De Historia Piscium* (1686), served as the ichthyological benchmark for more than sixty years—spelled out the

[34] The notebooks survive in the Long Island Collection of the East Hampton Library (KP 30 Mitchell [*sic*], Samuel Latham; MS. Lectures on Zoology). I cite here from volume 2. My thanks to John Walden, Dorothy King, and Marci Vail for their help with this collection, which has not been catalogued in any manuscript database (a mention of the books is made in Hall, but there is no evidence that he used them). Internal evidence gives one solid date in connection with the notebooks, 1817.

[35] The insertion "e" represents Mitchill's own labeled marginal annotation to his original text.

problem in 1693 in his *Synopsis Methodica*. Even though Ray recognized a fundamental distinction between pulmonary respiration and the respiration by gills—and saw (as Aristotle had) that this would seem to rank whales and dolphins with red-blooded terrestrial animals—he nevertheless stood by the ordering of *De Historia Piscium*, which included "cetaceous fishes" as one of the primary divisions of the class of fish:

> In order not to place ourselves at too great a distance from common opinion, and to avoid charges of gratuitous innovation, we shall classify the cetacean genus of aquatic creatures together with the fish,—even though they appear to correspond with the viviparous quadrupeds on essentially all points, except for hair, feet, and the element in which they live.[36]

Indeed, this very problem had helped give shape to the definition of fish in *De Historia Piscium*, where gills were not mentioned: "Aquatic animal, lacking feet, with naked skin or scales, swimming by means of fins, living perpetually in the water, and never of its own volition coming out onto dry land."[37]

What is seldom appreciated is that this ordering—of whales as fish—endured through the first nine editions of the presiding taxonomic authority of the eighteenth century, Linnaeus's *Systema Naturae*. Thus as late as 1756, Linnaeus—who since 1744 had defined breasts as "the essential characteristic of the 'Quadrupeds'"; and who knew very well that the cetes, as well as the manatees, gave suck to their young—remained committed to an order of cetaceans and manatees (collectively, the "Plagiures") ranked as a subdivision of the fishes, though he too felt it necessary to explain this seemingly anomalous grouping:

> All the animals of this order more readily bring to mind
> the Quadrupeds than the Fish, when we consider their internal structure: lungs, respiration, breasts, feet, appendages, live birth, etc. all indicate as much. We attach them, nevertheless

---

[36] The original reads: "Nos nè à communi hominum opinione nimis recedamus; & ut affectatae novitatis notam evitemus. Cetaceum Aquatilium genus, quamvis cum Quadrupedibus viviparis in omnibus ferè praeterquam in pilis & pedibus & elemento in quo degunt, convenire videantur, Piscibus annumerabimus." John Ray, *Synopsis Methodica Animalium Quadrupedum et Serpentini Generis* (London: Southwell, 1693), p. 55.

[37] Francis Willughby and John Ray, *De Historia Piscium* (Oxford: Sheldon, 1686), p. 1. I used the Princeton copy, Ex 8879.975q. The original reads: "Animal aquatile, pedibus carens, vel squammis vel cute nuda contectum, pinnis natans, in aquis perpetuo degens, nec sponte unquam in siccum exiens."

to the Fishes by reason of their *habitus,* their medium, their swimming, etc., in order not to fall from Charybdis onto Scylla . . .[38]

The point was that there were a host of different kinds of affinities presented by nature, and the job of the classifier was to do as little violence as possible as he sorted things into classes. The virtue of a great naturalist was the *judgment* with which he deployed the brute rules of his method. Whether whales were fish was, in this critical period in the history of classification, emphatically not a problem of facts; it was a problem of weighing the relevance of different facts to taxonomic order. In keeping the cetes with the fish in all his early work, Linnaeus opted for a general view of the animals in their totality. There was a personal dimension, too: in doing so, Linneaus preserved the ichthyological ordering of his dearest friend, his Uppsala schoolmate and natural-historical interlocutor, Petrus Artedi, whose fish classifications, published posthumously with Linnaeus's help as *Ichtyologia sive Opera Omnia de Piscibus* (1738), were the only leavings of a life cut mysteriously short in a drowning accident in 1735.[39]

It appears that the first naturalist to pull the cetaceans out of the fish category was the French zoologist Jacques Brisson, in his 1756 work, *Le Règne Animal Divisé en IX Classes,* which did not, however, go so far as to sort them with the quadrupeds, but instead placed them in their own class at the same rank, and made them the second of his nine major divisions of the animal world. When Linnaeus, in his tenth edition of 1758, declared that the cetaceans had to be detached from the fish and added to the mammals, "by good right and just title, according to the law of nature," Brisson complained that Linnaeus had gone too far along a path that Brisson himself had originally staked out. Indeed, he suggested that Linnaeus had only adopted such an exaggerated position (cetaceans-as-quadrupeds) so as to head off any allegation that he had plagiarized the central innovation (cetaceans-as-non-fish) from *Le Règne Animal Divisé en IX Classes;* Brisson, of course, believed the crafty Swede had done just that.[40]

---

[38] Linnaeus, *Systema Naturae* (Leiden: Haak, 1756), p. 39, note. Cited in Daudin, *Cuvier et Lamarck,* p. 125.

[39] On Linnaeus's friendship with Artedi, see Lisbet Koerner, *Linnaeus: Nature and Nation* (Cambridge, MA: Harvard University Press, 1999), pp. 34, 35, 154.

[40] See Jacques Brisson, *Ornithologie* (Paris: Bauche, 1760), vol. 1, p. xi. For the Linnaeus quote: *Systema Naturae* (Stockholm: Salvius, 1758), vol. 1, p. 16; see also the discussion in the editions of 1766–1767. For a review: Daudin, *Cuvier et Lamarck,* pp. 126–127.

That many systematic taxonomists gradually moved whales and dolphins out of the category of "fish" and into the category of "mammals" in the later eighteenth century was due in large part to the long shadow cast across natural history by the tenth edition of the *Systema Naturae* in those years. Hence, figures like Blumenbach and Camper ratified the move, and as Linnaeus's system was revised and disseminated by Gmelin and others, the non-fish whale grew to be an increasingly common feature of the naturalist's world. By 1789, when the Abbé Bonnaterre published the stand-alone quarto volume entitled *Cétologie*, the point of departure for the book would be a heated rejection of the position that whales were fish.[41] And about the same time (across the channel), the celebrated anatomist John Hunter—who had personally dissected a dozen cetaceans (mostly smaller ones), and who had engaged a surgeon on an English whaling vessel to collect specimens—could dismiss the question in his monograph on whales, announcing: "This order of animals has nothing peculiar to fish, except living in the same element, and being endowed with the same powers of progressive motion as those fish that are intended to move with a considerable velocity."[42]

But the very impatience of these assertions signals the durability of a contrary view (less well remembered in histories of classification), which survived above all among the followers of Buffon, for whom the Linnaean preoccupation with sorting natural productions into tidy boxes betrayed an overly simplistic view of the plural, plastic, and saturated character of nature. So, for instance, Duhamel du Monceau could, in 1782, include a discussion of the "cetaceous fishes" in the second part of his *Traité des Pêches*, and in 1792 (more than thirty years after Linnaeus's tenth edition) Vicq d'Azyr continued to follow Brisson in keeping mammals and cetaceans as separate classes at the same taxonomic rank. Buffon's aging collaborator Daubenton maintained the "viviparous quadrupeds" and the "cetaceans" as distinct classes as late as 1796, and even

---

[41] It is worth asking to what degree the firming of these positions in the period was bound up with the increasing specialization and institutionalization of zoology in these years. Or, to put it a different way, once there were experts in "cetology" (like Bonnaterre), the taxon had won itself a vigorous coterie of stakeholding defenders.

[42] John Hunter, "Observations on the Structure and Oeconomy of Whales," *Philosophical Transactions of the Royal Society of London* 77 (1787), pp. 371–450; Pierre Joseph Bonnaterre's *Cétologie* (Paris: Chez Panckoucke, 1789) was published as a separate book, but it was also properly a volume in the *Tableau Encyclopédique et Méthodique des Trois Règnes de la Nature*, the massive illustrated natural history published in Paris during the years of the Revolution (and continued well into the nineteenth century) under the editorial guidance of Charles-Joseph Panckoucke.

Lacépède—who spanned the era from Buffon to Cuvier at the veritable Mecca of global natural history, the Paris Muséum d'Histoire Naturelle, and who authored the major two-volume treatise *Histoire Naturelle des Cétacées* in 1803—felt obliged, in 1799, to distinguish "the quadrupeds properly understood . . . , from those animals that nature has, in reality, set aside by form, and even more so by habits, in such a way as to place them closer to birds and fish." The reference is to (fish-like) whales on the one hand, and to another problematic "edge" of the category of mammals, the (bird-like) bats, on the other.[43]

Cuvier's commitment to building a taxonomic system that privileged, above all, the differentia of comparative anatomy led him to class both whales and bats firmly within the Mammalia, and in his 1797 *Tableau Elémentaire*—a text, recall, that Mitchill kept to hand—Cuvier defended that arrangement. At the same time, he there demurred on following Linnaeus concerning a third perennially problematic "mammal"—man. In 1797 Cuvier continued to treat human beings as a distinct taxon, not appropriately positioned within the *mammifères;* only later would the king-maker of French science countenance a mammalian humanity. Mitchill would have learned of Cuvier's changed position on this contentious matter by the early years of the nineteenth century from Duméril's 1806 *Zoologie Analytique,* but it would not have shocked him, since Mitchill was already very familiar with Gmelin's edition of Linnaeus's *Systema Naturae,* which placed men and women firmly within the category of the "sucklers." Mitchill's own lectures in natural history in fact stayed closer to the Linnaean arrangement, presenting seven orders of Mammalia: from "*Cetae*" through "*Primates*—including man." It was in this sense that Mitchill could say—and mean, from the perspective of the taxonomy he taught—that "a whale is no more a fish than a man." The lawyers for the plaintiff would attack the "new philosophy" at this seam.

What, then, did Mitchill himself *really* know about whales in 1818? His books clearly indicated that the "new zoology" coming out of Paris embedded these creatures firmly in the category of the mammals, buttressing Linnaeus's final wishes, but Mitchill also understood perfectly well that this was a relatively recent, and not uncontested, position. For instance, returning to the list of reference works Mitchill used in writing his "Fishes of New-York" we discover that, of the four major treatises he had to hand, two kept some or all whales in the fish category: Laurentius Theodorus Gronovius, whose father had been Linnaeus's

---

[43] Daudin, *Cuvier et Lamarck,* p. 129.

patron, followed the earlier editions of the *Systema Naturae* by placing all the "Plagiures" under "Pisces"; Marcus Elieser Bloch was dead by the time René Castel prepared (under Bloch's name) the ten-volume 1809 Paris edition of the *Histoire Naturelle des Poissons*, and therefore was not in a position to object—as he might well have done—to Castel's decision to treat whales there as fish.[44] In fact, Castel even went so far as to refer to the front fins on a cetacean as its *"nageoires branchiales"*—i.e., its *gill*-fins![45] What Castel's volumes demonstrate is that authors and editors publishing market-oriented natural history about fish in this era had a hard time letting go of the whales: putting aside anatomy and physiology, these creatures were, in a way, the best part of the whole enterprise; and, anyway, most book-buyers expected to see them covered. Who wanted to disappoint the customer?

But the issue reached deeper than that: if Mitchill happened to use Wailly's French dictionary to help him through *Histoire Naturelle des Poissons* (it was a standard French vocabulary and grammar, easily available in New York City), he would have discovered the word "Cétacée" glossed as follows—"Said of large fish"![46] Now it is true that Mitchill's copy of Shaw's multivolume *Zoology* declared unequivocally that "the Cetaceous Animals, or Whales, however nearly approximated to Fishes by external form, and residence in the waters, are in reality to be considered as aquatic Mammalia." Yet the same text did subtitle the whole order "Fish-formed Mammalia," and acknowledged that "so strongly is the vulgar or popular idea respecting these animals impressed on the mind, that to this hour they are considered as Fishes by the mass of mankind."[47] Even in his revised second edition of the *Règne Animal*, published as late as 1829, Cuvier could still refer to whales as "the 'hot-blooded fish' of the ancients,"

---

[44] Bloch himself, in earlier editions, excluded the "great whales" from his ichthyology, but did include a number of the smaller cetaceans.

[45] Much to the infuriation of Adrien-Gilles Camper, who singled Castel out for abuse in the publication of Camper *père*'s posthumous volume on cetaceans: Pierre Camper, *Observations Anatomiques sur la Structure Intérieure et le Squelette de Plusieurs Espèces de Cétacés*, with notes by Georges Cuvier (Paris: Gabriel Dufour, 1820). See p. 21.

[46] "Se dit des grands poissons." For evidence that Noël François de Wailly's *Nouveau Vocabulaire François* (Paris: Rémont, 1803) was for sale in New York in this period, see *Evening Post*, 18 January 1806, p. 4.

[47] George Shaw, *General Zoology, or Systematic Natural History* (London: G. Kearsley, 1800–1826), vol. 2, pp. 471–472. At the same time, Shaw conveniently reprinted as an appendix to his second volume on mammals nearly the whole of Hunter's 1787 monograph on the whales from the *Philosophical Transactions* (Hunter, "Observations on the Structure and Oeconomy of Whales") so Mitchill would have had that work—which was very firm that whales were not fish—to hand (though without its detailed plates).

and he himself continued to use the outdated term "quadrupède" liberally through the 1820s.[48]

But Mitchill had other, less bookish, ways of learning about whales and dolphins too. As we have seen, he took students to the natural history cabinets of the city to pontificate about ocean fauna, and presented impromptu lectures on cetology while standing under the jaws of a baleen whale on Broadway. Moreover, where whales were concerned Mitchill did not fail to deploy his vaunted program for an inquisitive, collaborative, practical, and circulatory natural history. Mitchill was forever haunting the fish markets for natural-historical intelligence, some of which, he hoped, would have broader commercial value; these forays brought him into contact with whalemen, and others who preoccupied themselves with the cetes. For instance, as early as 1792 Mitchill corresponded with the Albany-based "Society for the Promotion of Useful Arts" concerning a method for catching porpoises that was used in New York waters, and he also detailed the process for tanning their skins into a workable leather.[49] Given Mitchill's expertise in fish and fisheries, his political activity, episodes of residence in Washington, and friendship with Thomas Jefferson, he was surely familiar with the exhaustive study on the cod and whale fisheries commissioned by Jefferson as Secretary of State and presented to the House of Representatives in 1791.[50] This document underlined the function of the whaling and fishing industries as schools for seamen, and as peacetime reservoirs for naval service, and went on to make a number of recommendations for how whalers and codfishermen might benefit from protective tariffs and other government measures. Mitchill's marriage into a shipbuilding fortune in 1799 increased his interests in maritime matters. He turned his attention to the causes of scurvy and to shipboard hygiene, and studied fish and fisheries with still greater focus.[51] He circumnavigated Long Island as part

[48]For "hot-blooded fish," see Cuvier, *Le Règne Animal* (Paris: Déterville, 1829), vol. 1, p. 69. On Cuvier's attachment to the term "quadruped," see Goulven Laurent, *Paléontologie et Evolution en France de 1800 à 1860: Une Histoire des Idées de Cuvier et Lamarck à Darwin* (Paris: Editions du C.T.H.S., 1987), pp. 76 and 101.

[49]Mentioned in John C. Greene, *American Science*, p. 92. Unfortunately, however, the reference he gives there (*Transactions of the Society Instituted in the State of New-York for the Promotion of Agriculture, Arts, and Manufactures* 1 [1792], pp. ix–x) does not appear to be correct.

[50]Thomas Jefferson, *Report of the Secretary of State on the subject of the Cod and Whale Fisheries, Made to the House of Representatives, 1 February 1791* (Philadelphia: Francis Childe and John Swaine, 1791). This was also reprinted in the House of Representatives Miscellaneous Documents (no. 32) for the 42nd Congress, 2nd session.

[51]Publishing, for instance, essays on the "Cod Fishery in the United States" and on the "Extension of the Means of Relief to Fredish Seamen" in 1804–1805. Both appeared in *The Medical Repository* 8 (1804–1805).

of his self-appointed mission of resource assessment, and promoted the consumption of fish, turtles, and even sea-elephant tongue, which he served to an early meeting of the Lyceum of Natural History. This oddment gives choice evidence of his links to the New England boats that were prosecuting the hunt for "elephant oil" at the remote south Atlantic ice limits in these years.[52] Sealers and whalemen were thus among his most important and far-flung informants, and in discussing the prolific *Clupea menhaden* in his "Fishes of New-York" (menhaden were a common baitfish in local waters), Mitchill reported his conversations with whalers, who had informed him that "on cutting up whales after death, great quantities of Menhaden had been discovered . . . in the stomach and intestine."[53] In fact, it appears that he went so far as to solicit from one of these old whalemen, Captain Valentine Barnard (from Hudson, New York), a pair of whale drawings: a sperm and a right [Plates 6 and 7].[54]

A review of the Minutes of the Lyceum of Natural History for the period leading up to the trial reveals that Mitchill and his fellow naturalists found themselves engaging cetological matters with some regularity.[55] In October of 1818, Mitchill conveyed to the cabinet "the jaws of the *Delphinus phocaena* or porpoise" sent to him by a New York shipping agent who worked out of South American ports, as well as the "fetus of a porpoise from the East Indies" sent along by another correspondent.[56] In the same year Mitchill would present a report on ambergris, "strongly corroborating the opinion that ambergrisse is nothing more than the

[52]On the trade in sea-elephant products, see Briton Cooper Busch, "Elephants and Whales: New London and Desolation, 1840–1900," *American Neptune* 40, no. 2 (1980), pp. 117–126. On the boiled sea-elephant tongue, see the manuscript Minutes of the Lyceum of Natural History for 12 May 1817, collection of the New York Academy of Sciences. On the turtle and fish, see Hall, *A Scientist in the Early Republic*, pp. 10–12, and Francis, *Reminiscences of Samuel Latham Mitchill*, p. 20.

[53]Mitchill, "The Fishes of New-York," p. 453.

[54]These drawings survive in the Issachar Cozzens Collection at the New-York Historical Society. While a good deal of this material was collected by Cozzens himself, quite a few items (particularly the earlier pieces) came originally from "among the rest of Dr. Saml L. Mitchill papers in the old poorhouse cellar [i.e., the basement of the New-York Institution]," as Cozzens put it in an annotation on item No. 87. I think the odds are excellent that the Barnard drawings came to Cozzens via Mitchill.

[55]Maritime studies were considered the organization's greatest strength, in large part (one supposes) because of Mitchill's leadership: see the comment by John W. Francis (a former member) in his 1858 book *Old New York*, where the Lyceum is briefly mentioned with the aside that "many of the rarest treasures of our marine waters have become known by the investigations of the Lyceum" (p. 372).

[56]Manuscript Minutes of the Lyceum of Natural History for 13 October 1817, collection of the New York Academy of Sciences. The correspondent was Jeremy Robinson (see *supra*, chapter 1, n. 3), some of whose letters (including correspondence with Mitchill) are preserved in the Library of Congress in the Peter Force papers.

indurated faeces of the *Physter macrocephalus* or sperm whale, and that it is never found in the healthy animal."[57]

Sperm whales were a topic of particular interest in 1818 at the Lyceum. In that year a committee was formed, consisting of four of the most expert zoologists in the group, to investigate a mysterious specimen sent to the society for identification: a giant tooth, thought to come from a "marine animal."[58] Such a specimen merited particularly close attention in early 1818, when the whole of the Atlantic coast was in the throes of the notorious 1817–1819 Gloucester sea serpent craze. Chandos Michael Brown has reconstructed the national and international reverberations set off by the "sightings" of an unusual sea creature gamboling in the waters near Cape Ann, Massachusetts in these years, and has examined the way that the authority of both lay witnesses and ostensible "experts" was challenged in the process of parsing fact from fiction in these incidents.[59] Mitchill and the Lyceum were swept up in the furor, and Mitchill himself deployed both his classical and his ichthyological erudition in an effort to contribute to a resolution. Shortly after the first reports hit New York, Mitchill read to the Lyceum a translation from the parts of Aldrovandi's work on fish (presumably the *De Piscibus* of 1613) "in which he describes several species of marine serpents," and Mitchill would later "debunk" a specimen brought forward as a sample of the creature's skin.[60]

While the committee assigned to work on the tooth returned that it was found to be the tooth of a "spermaceti whale," this was not wholly irrelevant to the pressing public mystery of the sea serpent, since sperm whales were known to feed on very large "sepia" or squid, and one of the favored explanations for the Gloucester creature among naturalists (in-

[57]Manuscript Minutes of the Lyceum of Natural History for 25 May 1818, collection of the New York Academy of Sciences. For a very interesting history of scientific investigation of ambergris, see K. H. Dannenfeldt, "Ambergris: The Search for Its Origin," *Isis* 73 (1982), pp. 382–397. By the early 1820s Mitchill was in fact using spermaceti as a specimen of "Materia Medica" in his lecture course at the College of Physicians. Whether he was doing so in the late teens I have been unable to establish. See College of Physicians and Surgeons, *Articles Contained in the Museum of Samuel L. Mitchill.*

[58]Manuscript Minutes of the Lyceum of Natural History for 9 March 1818, collection of the New York Academy of Sciences.

[59]Chandos Michael Brown, "A Natural History of the Gloucester Sea Serpent: Knowledge, Power, and the Culture of Science in Antebellum America," *American Quarterly* 42, no. 3 (1990), pp. 402–436.

[60]Manuscript Minutes of the Lyceum of Natural History for 24 November 1817, collection of the New York Academy of Sciences. Linguistic purists might like an admission here that the term "bunk" did not exist in 1817 (it was coined in 1820 in debates over the Missouri Compromise); "debunking" came almost a century later.

APPEARANCE OF THE (supposed) *SEPIA OCTOPUS* IN RELATION TO A SHIP AND A WHALE.

FIGURE 3. Mystery Creature of the Sea: woodcut that accompanied Mitchill's article "Facts and Observations Showing the Existence of Large Animals in the Ocean"; the bulbous monster looming up behind the ship could not be identified at the time (see text for an explanation). Courtesy of the Historical Library of the Cushing/Whitney Medical Library, Yale University.

cluding Mitchill) was that it was just such a giant "kraken." Mitchill published several reports related to "the existence of the Kraken in the ocean" in 1817, including correspondence from eyewitnesses (invariably ship captains).[61] This was a continuation of Mitchill's interest in large and mysterious sea creatures, an interest predating the Gloucester sightings, and that had resulted in a lecture published in 1814 in the *Medical Repository* entitled "Facts and Observations Showing the Existence of Large Animals in the Ocean, Different in Their Shapes and Manners from Whales, and Frequently Exceeding Whales in Magnitude," an essay accompanied by an illustration purporting to depict, to scale, a "(supposed) sepia octopus" [Figure 3].[62]

[61] See *American Magazine and Critical Review* 1 (1817), pp. 443–444 and the references there.

[62] This image is instructive, in that it indicates how unfamiliar even seagoing Americans were with the large rorquals (the general term for the baleen whales with grooved, expandible throats—for instance, what we now call blue and fin whales), which were not hunted in these years, as I discuss below, chapter 4. Any twentieth-century whaler would immediately identify

Mitchill thus had a recognized reputation as a seeker after the monsters of the deep, and after monstrosity in general, both by sea and by land: one of the first undertakings of the new Lyceum was to raise funds for the exhumation of a mammoth skeleton that had come to light in the loamy wet soil of an upstate farm. For Mitchill such a specimen would not only put another arrow in the quiver of all American naturalists who wished to lay to rest Buffon's lingering disparagement of the vigor of New World nature; it would also put New York on the map as a site for natural wonders by breaking Peale's monopoly on the monster, and, at the same time, it would let Mitchill enter more fully into debates about organic and terrestrial change in the region. It is also likely that Mitchill knew of the hypothesis put forward by Governor Thomas Pownall that the mammoth had been, in fact, a marine animal, and that "there are parts in the debris of the skull which have some comparative resemblance to the whale as to the purpose of breathing under water."[63]

Surely part of the public fascination with the trial of *Maurice v. Judd* can be linked to the timely broader interest in mysterious creatures of the sea occasioned by the Gloucester craze and the murmurs swirling around the recent excavations of giant bones—possibly marine—by naturalist-collectors in Philadelphia and New York. Mitchill's expertise in such matters was yet another qualification recommending his testimony. And at the same time, the perilous domains of mystery and monstrosity would afford another weak point in the edifice of classification, a weak point skillfully exploited by the opposing counsel.

Taxonomy at the Bar

After invoking the distributed natural-historical expertise of maritime New York, and then hazarding a novel natural-historical interpretation of Genesis, Samuel Latham Mitchill moved to the heart of his scientific expertise, and confronted the court with a version of his *Cetae* lecture

---

the looming, striated thing in this image as a dead rorqual floating upside-down in the water (so that its pleated underbody, swollen with the gasses of decomposition, arched up above the surface). The sight was a commonplace at the Antarctic whaling stations before World War II. Mitchill's article, and the illustration, appear in Mitchill, "Facts and Observations Showing the Existence of Large Animals in the Ocean," *The Medical Repository* 16 (1812–1813), pp. 396–407.

[63] DeWitt Clinton, "An Introductory Discourse," p. 106, note n. For an early suggestion that the mammoth was "of an amphibious nature," see Rembrandt Peale, "A Short Account of the Mammoth," *Philosophical Magazine* 14 (1803), pp. 162–169. Cohen (*The Fate of the Mammoth*, p. 46) notes that Leibniz identified a mammoth tooth as that of a marine animal.

from the College of Physicians and Surgeons: whales are not fish because they breathe air, have no gills, give live birth, and have horizontal tails. On these last points the piscatorial exceptions (dogfish, blennies, flounder) could be explained away by more thorough and refined familiarity with the animal world. Having thus sketched the classificatory boundary lines, Mitchill began to press deeper, asking his listeners to delve into the interior of the organisms, and deploy the techniques of proper comparative anatomy:

> The next distinguishing character is a vital mark; it is the structure of the heart: fish have but one ventricle and one auricle; but the cetaceous tribe have the heart double, like men and quadrupeds, and, as a consequence of this, alternate respiration.[64]

Other elements of the hidden internal anatomy were equally instructive: whales had a liver like that of a bullock, the doctor explained; unlike fish livers, the hepatic tissue of the cetes, exposed to air, rotted, and did not melt away into a usable oil.

William Sampson, probing for the cracks in Mitchill's testimony, tried to draw him onto the matter of the similarity of external form between the fish and the whales by noting that "Fins . . . are common to them both," but Mitchill used the scalpel of a comparative anatomist to ward off the suggestion, explaining that *under the skin* the "fins of the whale are of a peculiar structure," more like hands and arms. Sampson played for the laugh, while displaying that he had done his anatomical homework:

> Q: If, then, they are provided with hands and arms, it is natural to expect fingers and thumbs. How is it as to *carpus, metacarpus, and phalanges;* are they *present;* if so, could they use them for ordinary purposes, as to thread a needle, or do this? (taking a pinch of snuff.)[65]

To which Mitchill replied that "these extremities were covered with a membrane or web." Sampson, a skilled orator and a lawyer with a fearsome reputation in cross-examination, would not let up: "Like people

---

[64] *IWF,* p. 30.
[65] Ibid., p. 31; emphasis in original.

that wear mittens. No wonder they are awkward, and all their fingers like thumbs, as the saying goes."[66]

Such banter only set the stage, however, for Sampson's more elaborate attack on the value of taxonomic testimony. Putting aside a contest over anatomical facts (a tournament he was sure to lose), Sampson unfolded a two-pronged attack on Mitchill and the classificatory enterprise that he had come to court both to represent and to justify. In the first place Sampson—who was himself a member, like Anthon, of both the Literary and Philosophical Society and the Historical Society, and who was thereby in a position to marshal the societies' libraries of learned texts in the service of his advocacy—took Mitchill through the history of natural classification, emphasizing at every turn that dissent, discord, and change had long characterized the enterprise. Along the way Sampson also sifted out the continued existence of the Buffonian tradition which had never abandoned the category of the "quadrupeds" and which had never accepted the whales within that category. Here Sampson and Anthon were, in a way, borrowing a page from Mitchill's own book: as I have shown, Mitchill went to great lengths in his lectures and writing to emphasize that natural history was a dynamic enterprise in which dissent and argument were essential; Anthon, as it happened, had actually been Mitchill's student, and was himself a member of the Lyceum of Natural History. Coming in this way, from the inside, an attack that exposed the divided, contentious, even fickle nature of classification could be very damaging to the authority of Mitchill's scientific testimony.

Sampson's second line of attack was one that pointed to implications: If the "modern naturalists" (as both he and Mitchill called them) were offering a novel account of natural order, what did this mean for the place of human beings in nature? Were these new systems "better" than the old ones? And not merely "better" at sorting various creepy-crawlies. As Sampson would intone in his closing arguments:

> I think the onus lies on the advocates of this new philosophy to show to what good it tends. If it be to elevate the brutes, it is well contrived; but if it is for the benefit of the human kind, let them show what its virtue is. If it makes us better, happier, or

[66] Ibid., p. 31. Sampson's showy rhetorical style, and his particular success in the New York courts (which developed a distinctive character in this period, more open to grandstanding than the courts of Philadelphia and Boston), are discussed in Millender, "Transformation of the American Criminal Trial," chapter 3.

wiser, diminishes our toils, lessens our sorrows, or exalts our hopes, it is worthy of our gratitude and praise.[67]

This was a tall order for Cuvier, Lamarck, and Linnaeus, but these were questions that a democratic community might ask of its philosophers, particularly when the foundations of scripture and plain language were being catacombed by men seemingly obsessed with bones and bowels.

Both prongs of Sampson's attack—classification as a chaos, and the dangerous implications of the new philosophies of taxonomy—went after the cultural authority of the sciences in general and natural history in particular: at stake was not merely a few barrels of oil, or an ambiguously worded statute; at stake, in the end, was the proper place of science and men of science in the Republic, and, by implication, the civic function and standing of the three-story building just outside the courtroom, the New-York Institution of Learned and Scientific Establishments, so recently elevated as the city's distinguished Academy to stand significantly beside its new marble Acropolis.

Sampson's first line of questioning, then, worked to recover the long history of classification, and to excavate lost authorities in zoology, with an eye toward revealing that taxonomy was in fact a farrago of conflicting opinions. Take, for instance, the following exchange:

> COUNSEL: Have you present in your recollection, Doctor, the
> ninth book of Pliny's *Historia Mundi* . . .
> WITNESS: I have looked into Pliny, but I cannot trust to my
> memory.
> COUNSEL: Does it not appear from the writings of Pliny, that he
> knew the nature of the whales as well as the philoso-
> phers of this day?
> WITNESS: I do not think he did.
> COUNSEL: As to its mode of breathing and organs of respiration?
> WITNESS: I think anybody might know that.
> COUNSEL: Yet neither he nor Aristotle, nor any of the ancients,
> ever inferred from thence that the whale was not a fish.

To which Mitchill replied dismissively, "Very possibly, they were not so fully informed."

But was this high-handedness warranted? Sampson thought not: "I

[67] *IWF,* p. 73.

do not find that the modern philosophers agree as respects the system of Linnaeus; almost every one rejects some part, and substitutes or adds something of his own." A contention Mitchill answered glibly, in a manner not calculated to maintain the distinguished and aloof authority of zoologists: "They are, then, the more like lawyers, if they do not agree."[68]

This, of course, was precisely Sampson's point, and to make it palpable to the jury he led Mitchill through a painstaking recitation of the primary taxonomic divisions made by Aristotle (with and without blood), Pliny (*Homo, Terrestria, Aquatilia,* and *Insecta*), and finally Linnaeus himself, who, Mitchill declared, "first understood classification."[69] And though Mitchill complained in exasperation, "I must now appeal to the court, whether I am to be catechised and questioned like a college candidate," the judge gave Sampson the ground, explaining that "though a witness of superior learning or skill is called upon to speak as to matters of opinion, the other party is permitted to inquire into the grounds of that opinion."

Mitchill was thus obliged to nod as Sampson rehearsed Linnaeus's classes (*Mammalia, Aves, Pisces, Amphibia, Insecta,* and *Vermes*), then opened the mammals out into Linnaeus's seven orders (*Primates, Bruta, Ferae, Glires, Pecora, Belluae,* and *Cetae*), and finally laid the groundwork for his closing arguments by unfolding the *Primates* into Linnaeus's four genera ("*homo, simia, lemur,* and *vespertilio;* that is, man, monkey, macauco, and bat"). After Mitchill had affirmed these groupings as properly Linnaean, Sampson could begin to push on the larger implications of this taxonomy: "Now, is not man strangely mated or matched, when the whale and the porpoise are his second cousins, and the monkey and the bat his germans? Other gentlemen may choose their company, I am determined to cut the connection."

Leaving that looming question hanging, Sampson then turned to a skillful unraveling of the whole Linnaean enterprise, focusing on the monstrosity much in everyone's mind in the years of the Gloucester craze: the giant squid. Citing Mitchill's own writings on the kraken, Sampson sought further systematic guidance on these mysterious creatures:

> If I have not forgot what I once read in your Medical Repository touching this animal, it amounts to this: That in the first six editions of Linnaeus's systema naturae, it was placed amongst the *mollusca,* till Bomare and Banks rejected it altogether as fabulous. That *Ulisses Aldrovandus* and *Ambrosinus,* were, if I

[68] Ibid., p. 33.
[69] Ibid., p. 34.

recollect your very phrase, awed to silence, but that Mumfort came forward with proofs of its existence.

And when Mitchill confirmed this account, Sampson laid a trap of a question:

> The only thing I shall now request from your indulgence, is to know to what class, order, and genus, this prodigious creature properly belongs, according to the modern classification; if it be the same that Pliny described as being four acres in extent, *quaternum jugerum,* and is the same as the kraken, which is said to be a *cuttle fish,* and which Linnaeus classes with the worms, vermes.

Mitchill then demurred, explaining that the giant animal was nothing other than a very large instance of the "little fish [*sic!*] that comes to our market, called the squid."[70] But Sampson pushed the point: would Linnaeus ask that this mysterious and terrifying animal, larger than a whale, be classified with the *garden worm?* And was it not the case that at different times this same creature had been classified not only with the worms, but with the "oysters," and later still with the "crustacea," which would "necessarily associate it with the lobster, the scorpion, the pediculus or louse, and likewise with the flea"? Was it possible that modern zoologists had considered classing a sea creature with tentacles the size of a ship's mast in the very same order with infinitesimal black hoppers such as might be found on a *common dog?*

At this point the order in the court degenerated, since a number of spectators had been "seized with an impulse of laughter" at these incongruous groupings, and at Sampson's deft and theatrical assault on taxonomic pretense. Infuriated, the defendant's counsel (who had put Mitchill on the stand) rose to object, pleading with the judge to cut off "this mode of examining a witness of Doctor Mitchill's high standing and dignity of character." And out of this tumult, Mitchill offered a reply, not on the systematic standing of the sepia, but on the very idea of scientific progress. His peroration was abstracted by the shorthander:

> The witness then observed . . . that . . . generally speaking, the latest works of science were the best, as embodying the informa-

[70]Ibid., p. 37. Mitchill himself apparently here recurred to the "common usage" meaning of "fish" (a creature that lives in the water), since he himself was fully committed to placing squid in the order "Mollusca," concerning which he considered himself something of an expert. See Mitchill, "Facts and Observations."

tion of all predecessors, and leading to a correct summary of all
prior information; and he that would at this day trust to the
classical authors on zoology would do as one who, pursuing the
longitude, would trust to Kepler, and the mathematicians that
preceded Lord Napier.[71]

In high dudgeon, defending the onerous labors and considerable accomplishments of modern zoology, Mitchill continued:

> The difficulties of classification are great, requiring judgement,
> research, patience, and acumen. The necessities and difficulties
> are alike known to scientific men, and these difficulties are
> not aided by obscure references to the obsolete and antiquated
> doctrines. . . . The labors of necessary classification are to be
> painfully pursued and lead to infinite, and sometimes to micro-
> scopic researches, even amongst insects filthy to the touch, and
> disgusting to the sense.

But the hard-won results were among the crowning accomplishments of
mankind:

> Without methodical classification, and appropriate nomencla-
> ture, the study of the vast variety of nature's productions would
> be barren and impracticable. The great Linnaeus knew this, and,
> like a second creator, brought order out of chaos. His system is
> not perfect, neither can it be said to have been abolished, any
> more than the altering [of] this hall by shifting the benches, or
> changing the doors would be the destruction of the fabric.

And with that rousing declaration, Mitchill took his leave.

The issues that had been raised, however, would not go away. Seeking further to undermine both the defense's most illustrious witness and
the strongest part of their case, Sampson assembled a stack of books on the
plaintiff's side table, and the second day of the trial opened with Sampson's reading into the record the full weight of taxonomic authority that
stood against the zoological line that ran from Linnaeus through Cuvier
and Lamarck up to Mitchill. What Sampson had uncovered, reading in
the library of the New York Hospital in the days before the trial (and
in mining the relevant entries in several multivolume encyclopedias—
particularly the essay on "Cetology" in the *New Edinburgh Encyclopedia*),
was the durability of that Buffonian classificatory tradition, a tradition

[71] *IWF,* p. 38.

that expressed its anti-Linnaean bias through an entrenched opposition to "artificial" groupings. In the end, this tradition saw essentially all groupings as more or less the work of men, not nature.[72]

Citing the sentiments of "that great master genius," Buffon, whose work was "known to everyone" (via translations of the *Histoire Naturelle*), Sampson warned the jury that there were celebrated students of the natural world who had long protested against "the evils of burdensome nomenclatures, and tiresome minutiae." Within this Buffonian tradition, the Mosaical orderings were, after a fashion, preserved, and, to show as much, Sampson cited a number of English-language works in natural history all of which eschewed mammary-oriented neologisms, retained the older category of the "quadrupeds," and kept the cetaceans out of association with their terrestrial betters—for instance, Thomas Pennant's 1781 *History of Quadrupeds*.

This popular work (which ran quickly through multiple editions), by a fellow of the Royal Society who himself had an interest in cetacean anatomy, opened with the very sort of anti-Linnaean peroration from which Sampson cribbed his closing arguments: Pennant acknowledged his respect for the great Swede, but complained that the constantly changing editions had made the *Systema Naturae* difficult to defend as a "system"; worse still, the category of the Mammalia was, for Pennant, impossible to recommend, and he wrote that the faults in this class "oblige me to separate myself, in this one instance, from his crowd of votaries."[73] The cetaceans—along, again, with man ("my vanity will not suffer me to rank mankind with apes, monkeys, maucaucos and bats")—were the rub:

[72]For a review of the English debates around the Linnaean system in this period, see Mario A. DiGregorio, "In Search of the Natural System: Problems of Zoological Classification in Victorian Britain," *History and Philosophy of the Life Sciences* 4 (1982), pp. 225–254. Sampson includes a reference to the "Cetology" article (*IWF*, p. 83), and a reading of it leaves no doubt that it is from this source that he assembled the reference base he used to undermine Mitchill's testimony. See "Cetology," in David Brewster, ed., *New Edinburgh Encyclopedia*, 2nd American edition (New York: Whiting and Watson, 1814). It is interesting to think about the use of encyclopedias in the trial (they are referred to four times) in the context of Yeo's article on encyclopedias and professionalization in the sciences in this period: Richard Yeo, "Reading Encyclopedias: Science and the Organization of Knowledge in British Dictionaries of Arts and Sciences, 1730–1850," *Isis* 82, no. 1 (1991), pp. 24–49. See also his *Encyclopaedic Visions: Scientific Dictionaries and Enlightenment Culture* (Cambridge: Cambridge University Press, 2001), and Arthur Hughes, "Science in English Encyclopedias, 1704–1875," *Annals of Science* 7, no. 4 (1951), pp. 340–370.

[73]For Pennant's interest in cetaceans, see the "Description of the Blunt-Headed Cachalot" by James Robertson, which Pennant communicated to the Royal Society in 1770 (*Philosophical Transactions of the Royal Society* 60 [1770], pp. 321–324).

> The last order is that of whales: which it must be confessed, have, in many respects, the structure of land animals; but their want of hair and feet, their fish-like form, and their constant residence in the water, are arguments for separating them from this class, and forming them into another, independent of the rest.[74]

Pennant was, in a way, recovering the taxonomic value of the older grouping of four-footed beasts ("quadru-peds") exactly to defend against those three unseemly "edges" of the territory Linnaeus had defined as the Mammalia: whales, human beings, and bats. The Buffonian preference for what we would now call "family resemblance" categories—the "dog-like animals," "the cow-like animals"—was a powerful resort for a classifier who had watched the Linnaean system move further and further from "common sense" in the familiar realm of the beasts. And one could hold this view even while admitting that the author of the *Systema Naturae* had done extraordinary things sorting plants and the lower animals, where his achievement was nothing less (as Pennant put it) than to have "given philosophy a new language." Where the big critters were concerned, however, Pennant cast his lot with Linnaeus's archrival (though not without acknowledging some misgivings about Buffon's thoroughgoing unwillingness to "shackle his lively spirit with systematic arrangement"). How intimately was Pennant tangled up with the great French naturalist? So much so that the whole of Pennant's *History of Quadrupeds* was originally conceived as "an index to Buffon."

Pennant was by no means alone. Buffon had been blessed with a gaggle of English-language naturalists who busily redacted his work in the period, most prominently Oliver Goldsmith, whose 1774 rewrite of the original *Histoire Naturelle* into a piece of inspired hackwork (*A History of the Earth and Animated Nature*) achieved irrepressible popularity.[75] More than twenty nineteenth-century editions and digests of this multivolume work can be traced, testimony to the prescience of James Boswell, who commented that Goldsmith would make natural history "as entertaining as a Persian Tale."[76] Thanks to the source criti-

---

[74] Thomas Pennant, *History of Quadrupeds* (London: B. White, 1781), vol. 1, p. iv. The passage is unchanged (at pp. iv–v) in the 1793 edition, which is the one Sampson is most likely to have used.

[75] Porter discusses the movement of Buffon to the United States in this period. See Porter, *The Eagle's Nest*, p. 16.

[76] The remark is quoted in Donald M. Hassler, "Enlightenment Genres and Science Fiction: Belief and *Animated Nature* (1774)," *Extrapolation* 29, no. 4 (1988), pp. 322–329, at p. 325.

cism of James Hall Pitman we have a very clear sense of how Goldsmith cut and pasted from Buffon, stringing these excerpts together with pleasing anecdotes and amusing aperçus in an effort to compose a work indifferent to the pretensions of zoological "science," but concerned instead to mobilize natural-historical knowledge in the service of pleasurable reading: here was the "book of nature" as leisure, not labor. As Goldsmith himself put it: "my chief ambition is to grab up the obscure and gloomy learning of the cell to open inspection; to strip it from its garb of austerity, and to show the beauties of that form, which only the industrious and the inquisitive have been permitted to approach."[77]

Both sides in *Maurice v. Judd* had access to the 1816 London edition of this work (a six-volume version expanded by W. Turton, a Fellow of the Linnaean Society), and both read excerpts from it into the record of the trial. For Sampson, Anthon, and the plaintiff, Goldsmith was a valuable ally, since his popularity made him a familiar reference for many in the court, and his Buffonian natural history meant that *Animated Nature* had no patience for tedious anatomical details or counterintuitive groupings based on disgusting or louche features. Where the fishes were concerned, Goldsmith followed the classic formulation of Willughby and Ray, identifying three primary classes—*cetaceous, cartilaginous,* and *spinous* (or bony)—and did so with full knowledge of the urgings of the modern anatomists, which occasioned merely an aside on the limited value of comparative anatomy for classification:

> A different formation of the lungs, stomach, and intestines, a different manner of breathing or propagating, are not sufficient to counterbalance the great obvious analogy which these animals bear to the whole finny tribe. They are shaped as other fishes; they swim with fins; they are entirely naked, without hair; they live in the water, though they come up to breathe; they are only seen in the depths of the ocean, and never come upon shore but when forced thither. These sure are sufficient to plead in favour of the general denomination, and acquit mankind of error in ranking them with their lower companions of the deep.[78]

Goldsmith, like any good populist, cast his lot with the folk.

---

[77]Cited in James Hall Pitman, *Goldsmith's Animated Nature: A Study of Goldsmith* (New Haven: Yale University Press, 1924), p. 15.

[78]Oliver Goldsmith, *A History of the Earth and Animated Nature* (London: Wingrave and Collingwood, 1816), vol. 5, p. 27.

In addition to citing Pennant and Goldsmith, Sampson also brought forward as evidence slighter texts which were little more than dilapidations of these higher exercises in English book-trade natural history, works like John Bigland's epistolary rendering (or, more properly, piracy) of Goldsmith, which, under the title *Letters on Natural History* (companion volume to the *Letters on Universal History*), assembled sixty-two didactic missives on nature "calculated particularly for the use of schools and young persons in general of both sexes; in order to impress their minds with a just idea of its great author."[79] Bigland too, predictably, placed the whale at the head of the sea creatures, and informed his reader-penpal that this "enormous fish" must be "regarded as one of the greatest curiosities of animated nature; and if its commercial importance be justly appreciated, it will be esteemed an object worthy of the attention and examination both of the naturalist, the politician, and the merchant."[80]

Calling on these texts, Sampson proved to the jury that there was a durable, indeed prolific, body of natural history literature that held Linnaean categories at bay, and that preserved whales as the kings of the fish. By doing so Sampson succeeded in turning Mitchill's vision of a vigorous, intellectually dynamic world of natural-historical disputation against itself. Moreover, Sampson pursued this subversive attack along several vulnerable avenues. Not only did taxonomists disagree about where whales belonged in the order of nature, but also (and worse) the whole enterprise of natural history was, Sampson asserted, riven by internal contradiction: drawing on an established anti-Linnaean argument, this skilled debater set about undermining the whole contemporary project of classification by setting its preoccupation with novelty against its pretensions to discover immutable hierarchies. It was, he pointed out, emphatically the joy of the modern zoologists to proffer novel and heretofore undiscovered creatures at every turn: krakens and mammoths, duck-billed beavers and kangaroos. Given then that many new animals were constantly being added to the pantheon of natural history, how could any taxonomic system—of necessity based on an incomplete sample—pretend to have seized on the proper classifications for *everything?* Here Sampson was rehearsing a well-worn objection raised (by Buffon and others) against Linnaean-style rubrics for sorting the plenum into the nesting trays of some vast and idealized cabinet. Of

---

[79]I have consulted the second edition, 1810, in the holdings of the New York Public Library, but find no record of a first edition in any catalogued collection in the U.S. or abroad.

[80]John Bigland, *Letters on Natural History* (London: James Cundee, 1810), p. 340.

the perennial flood of novel animals Sampson asked: "had they been known," might they not have served as "connecting links, and prevented such grotesque associations and abrupt transitions," and might they not have helped avoid the "forced and incongruous grouping of animals every way dissimilar, in the same order; as in the order *ferae,* or wild beasts, the lion, the seal, the hedgehog and the *mole*"?[81] And if taxonomic systems were designed in such a way as to be always ready to admit of the novelty that natural historians so prized, then would not each system have to become a perpetual pageant of rearrangement? As Sampson put it, "the ranks must open to receive them as they respectively arrive, and the order of the parade be still changing."[82] How could the jury repose its confidence in an enterprise so at the mercy of whim and circumstance?

Indeed, in his closing arguments, Sampson stressed exactly this problem—the heterogeneity of zoological orderings—the issue with which he had battered Mitchill on the stand. Referring the court to the very long article on "Classification" in the American edition of Rees's *Cyclopedia* (a monograph-length entry that ran to several hundred pages and extended over two volumes in a work that Mitchill himself had cited, praised, and patronized), Sampson pointed out that those who would overturn Linnaeus were not some wayward rump of radicals, since after all "Linnaeus had overturned the systems of Artedi and his predecessors," and moreover "the names of the sectarians who have since founded other systems, or new modelled [*sic*] his, would fill a catalogue."[83] Equipped with further evidence that the history of classification was a hurly-burly and a perpetual game of king of the hill, Sampson asked the jury how anyone could possibly begin to find his way among the competing claimants. Pressing the point in a trenchant rhetorical overture, Sampson asked if there would be any end to the arcane and fantastical ratiocination of the modern systematists:

> And what virtue is there in the scalpel of the anatomist, or in his blow-pipe, that he should have the sole privilege of newcreating, and nicknaming God's creatures? Why may not the

[81] *IWF,* p. 35.

[82] Ibid., p. 70. For a period discussion of this problem in the literature of natural history, see Pennant, *History of Quadrupeds,* introduction.

[83] *IWF,* p. 42. On Mitchill's support for Rees, see the manuscript letter (dated 3 March 1820) seeking subvention of the publication of the American edition, signed by Mitchill (along with DeWitt Clinton, David Hosack, and others), which calls the *Cyclopedia* a "rich and valuable body of information." Archives of the Historical Society of Pennsylvania.

school of Boerhaave found a system of zoology upon the *ele-menta medicinae physico mathematicae,* and class according to the animal functions, explained by mechanical causes and mathematical demonstrations? [¶] The craniologist, craniosophist, cranionomist, or craniogonist, after Gall and Spurzheim, may by the nervous streaks on the ganglion of the brain, and their circumvolutions on the hemispheres, and the protrusions by the organs of the mental faculties of the parallel lamina of the skull . . . class and arrange all things.[84]

And, reflecting a familiarity with the anthropometry that was already an essential element of the classification of the human races, Sampson raised specters still more extravagant and troubling:

The geometricians will have pantometrical zoology, classing by points, lines, surfaces, and solids, the genders of curves, ratios of affections, motions, and positions. There will be the straight, crooked, and perpendicular families, the equilateral, curvilateral, and multicrural: The cycloid, trapezoid, and rhomboid orders, so that by comparing the angle of the bee's knee with the inclination of the cat's tail, there will be no danger of mistake. And with the aid of a Gunter's scale, a logarithmic table, and a theodolite, you may distinguish a maiden from a bat. [85]

This biting satire nibbled its way to the crux of the matter: the implications—for human beings—of the new taxonomies. It was here that the plaintiff's side saw their best hope for undermining not only Mitchill but the whole edifice of modern zoological classification.

Zeroing in on the Linnaean category of the *Primates,* Sampson linked the fate of mammalian whales to that of primate men and women, reminding the jury that if they accepted Mitchill's testimony on the cetes, they were obliged to accept as well a humble place for themselves in the order of nature: "Yes, gentlemen of the jury, in the same order with man,

[84]The question marks in this quote I have inserted; they appear as full stops in the original text. *IWF,* p. 70.

[85]*IWF,* pp. 70–71. See also p. 73, where Sampson pressed the issue of the maiden and the bat, playing off Linnaeus's early classification of the bats as belonging in the category "quadrumana": "figure to yourselves gentlemen, a beautiful woman in the act of nursing her first child, her eyes beaming with tenderness and love, and grace and beauty in her form and attitude; and now imagine me a Borneo bat, clinging torpidly in some filthy hole, by the hooks which terminate its hinder extremities, its head downwards, its ugly brood sticking to its dugs, and reconcile, if you can, this fantastical association to decency or reason."

they place the monkey, ape, and baboon; all equally related, and differing from the lord of the creation only as they differ from each other."[86] And, Sampson suggested, the Linnaean *differentia* were hardly comforting:

> But now, the rule is this, and if you follow it, you will be quite safe. A man is an ape *minus* two thumbs, and a baboon *minus* two thumbs and a tail. And *e converso,* a baboon is a man *plus* two thumbs and a tail, and a monkey is a man *plus* two thumbs; or thus, a man with an extra pair of thumbs would be an ape, and with those, and the addition of a tail, would be a baboon. If I had not known this to be philosophy, I should have supposed it was the *black art.* It is enough to give bad dreams.

Conveniently, Sampson's co-counsel, John Anthon, was there to conjure up those very dreams, darkening the nightmare with the bogies of race, civic disorder, and excessively universal franchise. Stirring up a witches' brew of Mitchill's modern taxonomy and entrenched American legal hierarchies, Anthon warned of what might lie ahead if the men of science were given license to interpret the statutes of the state. Reiterating that the implication of Mitchill's analysis was that the monkey is, "in the language of naturalists, no more a brute, than a whale is a fish," Anthon went on to ask how far Linnaean classification would be permitted to penetrate into the system of law:

> We have a statute, which declares that every freeman shall be entitled to a vote at our public elections; let us suppose, then, that at some one of those arduous struggles, where everything in the shape of a man had been by the zeal of politicians urged to the hustings, that the learned Doctor had appeared, leading forward with all due gravity to the polls an orang outang, or man of the woods, would the stranger's vote be received, although the Doctor should very learnedly and eloquently urge his claims, as he has done those of the whale on the present occasion? He breathes the vital air, the Doctor might say, through lungs; He moves erect, &c. has warm blood; the female brings forth her young alive, and rears the bantling at her breasts. The inspectors would say, in reply to all the eloquence and learning of the Doctor, as we do in the

---

[86] *IWF,* p. 74. For helpful context on American contact with and ideas about primates, see Brett Mizelle, "'Man Cannot Behold It Without Contemplating Himself': Monkeys, Apes and Human Identity in the Early American Republic," *Pennsylvania History: A Journal of Mid-Atlantic Studies* 66 (1999), pp. 144–173.

case of the whale, all this indeed is very strange and curious, but still, Doctor, it is a monkey in common acceptation, however naturalists may choose to hail and class him as a brother.[87]

No one in the Mayor's Court would have missed the import of these high-stakes histrionics: In 1818 New York was in the throes of a contentious political and cultural transition; just a year earlier the legislature in Albany had passed the controversial emancipation statute of 1817, which set in motion a series of steps that promised freedom for all the slaves in the state by 1827. This move initiated a period of increasing White hostility toward African Americans in the city, as fears of labor competition and social mingling precipitated violent reaffirmations of racial supremacy. These developments made the dangerous issue of the free Black franchise—a source of sharp and long-standing disputes between the Democratic-Republican party (with its ties to the South) and the Federalists (with whom free Blacks were thought to vote)—more explosive than ever. It had been widely alleged that the three-hundred free Black voters of New York City had been responsible for the Federalist victory in the closely contested Assembly election of 1813, and wartime anger among anti-Federalists spilled over against the Black community in the years that followed. Various legal and illegal means were deployed to harass and intimidate potential Black voters in the second decade of the nineteenth century, and the specter of complete emancipation had given new urgency to the promoters of such practices after 1817.[88]

All of which is to say there was no doubt in the room that Anthon's fear-mongering was intended to invoke the inflammatory matter of race and franchise in the city, tacitly playing on the pernicious and durable period rhetoric (and nascent race-science) that linked Africans to the lower primates (Anthon's "hail him as a brother" mischievously echoed the familiar slogan of the abolitionists, "Am I not a man, and a brother?").[89]

---

[87] *IWF*, p. 61.

[88] For a discussion of these events, see chapters 3 and 4 of Leslie M. Harris, *In the Shadow of Slavery: African Americans in New York City, 1626–1863* (Chicago: University of Chicago Press, 2003). Also useful: Graham Russell Hodges, *Root and Branch: African Americans in New York and East Jersey, 1613–1863* (Chapel Hill: University of North Carolina Press, 1999), particularly chapter 7. More generally on the changing party politics of the period: Dixon Ryan Fox, *The Decline of the Aristocracy in the Politics of New York, 1801–1840* (New York: Harper and Row, 1965); and Harvey Strum, "Property Qualifications and Voting Behavior in New York, 1807–1816," *Journal of the Early Republic* 1 (1981), pp. 347–371. For the role of the Clintonians, see Evan Cornog, *The Birth of Empire: DeWitt Clinton and the American Experience, 1769–1828* (New York: Oxford University Press, 1998), particularly pp. 138–144.

[89] For a detailed discussion of the texts and ideas that would have informed Anthon's listeners on this topic, see Bruce Dain, *A Hideous Monster of the Mind: American Race Theory in the*

Line-drawing and category-defining were the central problems of the franchise in these years; would the (White) citizenry care to hand that fraught business over to a clique of didactic, self-important professors? Interestingly, while both Sampson and Mitchill were involved in manumission activities, they had recently squared off in another legal confrontation, one that brought questions of medicine and taxonomy directly to bear on the issue of race. Back in August of 1808, Sampson and Mitchill took center stage in the sensational case of *The Commissioners of the Almshouse v. Alexander Whistelo,* in which the African-American man Alexander Whistelo (a coachman in the service of the botanist David Hosack) challenged the allegation that he fathered the child of Lucy Williams, "a yellow woman"; he thus disavowed financial responsibility for her offspring.[90] Sampson represented Whistelo in the case (which meant the Irish lawyer had a vested interest in nurturing the widespread rumors that the child was in fact Hosack's). Medical testimony dominated the trial, which turned on the question of whether the child could plausibly have been Whistelo's, given that "the said child was of quite a light color."[91] Mitchill (a close friend of Hosack's, a fact not overlooked by the chatterers) was one of very few medical men to testify that it was, in his opinion, "a possibility, nay, a probability, that the said child has been begotten by the said Alexander Whistelo."[92] The case solicited medical and scientific witnesses to discuss the taxonomy of racial admixtures, and the physiology of child development among the different races. Anthropometrical techniques of racial discrimination were discussed in this case, which was apparently sufficiently well known among nineteenth-century students of law in New York (primarily for its bordello scene testimony) that the third volume of Wheeler's *Reports of Criminal Law Cases,* where the transcripts originally appeared, proved very difficult to keep on law school shelves.[93]

Early Republic (Cambridge, MA: Harvard University Press, 2002), as well as Winthrop D. Jordan, *White over Black: American Attitudes toward the Negro, 1550–1812* (Chapel Hill: University of North Carolina Press, 1968).

[90]The case is discussed by Irving Browne in "William Sampson," *The Green Bag* 3, no. 8 (1896), pp. 313–325, at p. 324. I have read the proceedings in Jacob D. Wheeler, *Reports of Criminal Law Cases* (New York: Banks and Brothers, 1860), pp. 194–236. The quote comes from p. 194.

[91]Wheeler, *Reports of Criminal Law Cases,* p. 196.

[92]Ibid., p. 197.

[93]The publishing history of Wheeler's *Reports* is complex: my reference to the "third volume" indicates not the 1860 printing cited above, but rather the original series publication of 1823–1825. There is surprisingly little secondary literature on *The Commissioners of the Almshouse v. Alexander Whistelo,* given the interesting issues at play in the case. It is briefly discussed in Shane White, *Somewhat More Independent: The End of Slavery in New York City,*

The Commissioners of the Almshouse v. Alexander Whistelo demonstrated that Mitchill and his naturalist colleagues were quite happy to weigh in on the issue of racial categorization, using new technologies and new theories to rearrange boundaries and defy common sense. When Anthon, playing to the gallery in *Maurice v. Judd*, allowed himself a rhetorical dalliance with political bestiality in the coded language of race, he showed how much was at stake when the new philosophy of classification came to the bar.

From the outset, Samuel Latham Mitchill's trip to the witness stand in the case of *Maurice v. Judd* promised to stage a very public showdown, a duel of wit and erudition, between the city's leading naturalist-philosopher and a pair of its prominent attorneys. The occasion beckoned to the citizens of the metropolis in the leisurely days between Christmas and the New Year precisely because sparks were likely to fly when Sampson and Mitchill crossed their rhetorical blades in the Mayor's Court, and rumors of Mitchill's "paradox" ensured that the galleries were full when it came time for him to defend the learning of the French school in the teeth of common opinion. As we have seen, Mitchill worked to defend more than the taxonomic status of air-breathing marine creatures with horizontal tails: he tried to defend a vision of natural history that was collaborative and centripetal, that centered on the maritime city of New York, and that derived its intelligence from the expanding global orbits of Fredonian seamen. When flushed from this cover by the baying pack of democratic consensus (which howled that all sea creatures were, by definition, fish) Mitchill resorted to a pious exegesis of Genesis, and buttressed his natural history with a textually attentive natural theology well rooted in scripture. Only then, after allying himself with the intelligence of the Republic and a Mosaical zoology, did Mitchill rehearse the "principles of science" upon which the non-fish whale rested. As he put it to the court, according to these principles, "as now digested, perfectly understood, and past all question, the facts being all arranged and posted up to this day, and as far as human discoveries have gone, and

---

1770–1810 (Athens: University of Georgia Press, 1991), pp. 162–163. For further evidence of Mitchill's continued investigations of racial science (specifically the taxonomically and politically troublesome issue of darker-skinned people "turning white"), consider the article published in Trenton, New Jersey, "'The Ethiopian' has changed 'his skin,'" *Federalist,* 1 November 1824, p. 2.

human research penetrated, it is received as an incontestible [*sic*] fact in zoology, that a whale is no fish."[94]

As Sampson drew Mitchill's testimony firmly onto the terrain of modern taxonomy, he drew the doctor more deeply into a defense not merely of mammalian cetaceans, but of the very methods, results, and implications of contemporary natural history, calling into question the intestine preoccupations of comparative anatomy, the counterintuitive classifications of Linnaeus, Cuvier, and Lamarck, and the unsettling company that human beings were supposed to keep under these novel schemes. Along the way Sampson emphasized alternative taxonomies (like those of Buffon) that remained current and that rejected Mitchill's groupings. Even more damaging, Sampson marshaled a great deal of evidence that the whole history of taxonomy was nothing other than a chaotic procession of dissent and discord, a litany of failed systems and self-aggrandizing systematizers each discarding the work of his predecessors. I have suggested that here, in an effort to undermine natural historical authority, Sampson (and Anthon) were successfully mobilizing Mitchill's own depiction (in his teaching and writing) of natural history as a vigorous and dynamic enterprise, full of argument and dissent.

In this subversive and skeptical assault on the whole enterprise of natural history we garner a remarkably clear view of just how unsettled the domain of the classificatory sciences could look in a period that historians of science have tended to characterize as the triumphant age of taxonomy. A skilled and intelligent disputant, with access to a modest medical library, could scope out deep rifts in the program that Mitchill depicted as having achieved thoroughgoing consensus.[95] By drawing on the collatitious and piratical domain of book-trade natural history, and sifting the volumes of encyclopedias like Rees's and Brewster's, Sampson was able to assemble compelling evidence that taxonomy was a fraught and fallible enterprise: the very possibility of a single, presiding, fixed scheme for arranging the productions of nature remained in question, and competing accounts could be seen to cancel each other out; novelty,

[94] *IWF,* p. 25.

[95] It is worth noting that the kind of erudition that Sampson displayed during the trial— the product of a quick "cram" using encyclopedias and other reference works—had been for some time a source of anxiety (particularly in Britain) among "true" savants, who looked on in horror as a hyperactive book trade retailed "learning" at bargain-basement prices. For a helpful introduction to these disputes, see Roger D. Lund, "The Eel of Science: Index Learning, Scriblerian Satire, and the Rise of Information Culture," *Eighteenth-Century Life* 22, no. 2 (1998), pp. 18–42.

monstrosity, and humanity all proved weak points in the schemes of those who would collect and sort all the creatures of the globe; the vaunted claim of the naturalists—that they had peeked into God's own cabinet, and witnessed how He arranged His creations—could be made to sound like a terribly hollow boast. What remained after this assault was hardly a picture of human mastery over creation. Rather, here were the feeble offerings of those "who attempt to follow nature into the secret recesses where she loves to retire" and who thus "find themselves in a labyrinth to which they have no clue." Hence, as Sampson continued,

> those trembling lines, traced with a feeble hand; those limits which nature disavows, and boundaries which she disowns. Hence those ephemeral systems which efface each other, succeeding like the ocean's waves, of which the inventor has only this benefit, that his errors are concealed under the protective cloud of mystical jargon.[96]

Nor was Sampson content to have undermined the authority of zoologists and classifiers. He dilated his argument into a broad assault on the pretensions and dangers of science more generally, and permitted the dominant metaphors for scientific progress to undo themselves: Linnaeus, Mitchill would have it, brought forth order like a "second creation"? Sampson answered that we would do better to "remain as we were before this second creation, for I do not think it becoming for the lords of creation to be ranked with porpoises and hogs."[97] Was it the case that more recent science universally excelled that of the past, since each generation stood on the shoulders of the last (as Mitchill had suggested)? Sampson parried: "so unsteady is the footing of those who stand upon each others [*sic*] shoulders, that I fear this *cumulative series* will not gain strength by numbers, and if the most learned are uppermost, there is danger of the column becoming top heavy. It was in this manner the giants tried to climb, but their pride got a fall, and I fear this new philosophical ladder will scarcely reach where Jacob's did."[98] In his closing arguments Sampson would prove as much, since he would wrap himself (and his client) in scripture, showing that the English Bible used "great fish" to describe the creature that swallowed Jonah in the Old Testament, but that when

---

[96] *IWF,* p. 71.

[97] Ibid., p. 73.

[98] Ibid., p. 69. For a remarkable tracing of the history of the trope at issue here, see Robert K. Merton, *On the Shoulders of Giants: A Shandean Postscript* (New York: Free Press, 1965).

the Gospel of Matthew alluded to the same event, the reader learned that the reluctant prophet had been "three days and three nights in the *whale's* belly."[99] So much for Mitchill's Mosaical-Genesical distinction between *Pisces* and *Cetae.*[100]

And just as science itself came under suspicion in *Maurice v. Judd*, the persona of the man of science—and his place in the Republic—also came under assault in the closing arguments of the case. Likening Mitchill to an Aristotle aggrandized by the despotic power of that "Alexander" known as "public opinion," Sampson depicted the doctor as a fearsome force: "Who but one conscious of his terrible power would have planted himself in the attitude of proud defiance, like a castle on a cliff, and proclaimed in the face of a court and jury what he did, on that ever memorable moment, when he declared, upon the faith of modern zoologists, *that a whale was no more a fish than a man?* and that none but lawyers and politicians would now a days suppose it so?"[101] Here was a man of learning who was quite nearly a corrupter of youth, and Sampson threatened that he might just have to take a walk across town and visit Mitchill's lecture room, in order to inform his misguided acolytes that, *pace* their sophisticated opinions, the common law did indeed deem the whale very much a fish, and nothing whatsoever like a man. Centuries of legal tradition and case law buttressed the position. Again and again Sampson sounded the alarm at the "monstrosity" of the man of science loose in the court, and Sampson defended his rough handling of the esteemed witness by arguing that "the manner in which he has treated the law and the lawyers, considering his might and his weight, makes it as much self-defence, as it would be to turn an elephant out of your garden where he would trample everything under his feet."[102] Just as the whale was a strange sort of monster-hybrid—"from the midriff down it has no one of the character [*sic*] of man or quadruped, so that all the power of philosophy, with 'oxen and wain-ropes' to boot, can never get it more than half out of the water"—so, too, the Jeffersonian philosopher had about him something of the amphibian, "of the people" one moment, lecturing them the next, lording his erudition over them like a beast, and

---

[99] *IWF,* p. 73; emphasis original.

[100] It should be noted that there is much to be said about the etymological trajectories of the two Latin terms that come to be translated as "whale," and there are implications for biblical exegesis as well as the history of whale-knowledge. For a taste of all this, consider: James Byrne, "Cetus and Balena," unpublished paper, 2003.

[101] *IWF,* p. 69.

[102] Ibid., p. 38.

then vanishing into their bosoms.[103] Indeed, Sampson likened his combat with Mitchill to the encounter with the ultimate amphibian of myth:

> There was a demigod of old, a great icthyologist [*sic*], for he kept the flocks of Neptune. He knew almost every thing, and was consulted upon all great occasions, by the kings and people. He was Proteus, and could, if not disposed to answer categorically, use all shapes, shifts, and transitions. It was so with the learned witness in this cause; you saw that he eluded our inquiries; how he flew from the arctic to the antarctic circle; from Davis's straights to the straights [*sic*] of Magellan; and like Puck, the fairy, put a girdle round the earth in forty minutes.[104]

And this depiction of the man of science as a protean, formless creature—a nondescript, a beast neither fish nor flesh—ran throughout the trial, reaching perhaps its most elegant formulation in Sampson's cross-examination of Mitchill, when, in the course of questions about the amphibious nature of the manatee, Sampson drew Mitchill into an exchange about the great sea monster of classical antiquity, Oannes, "who daily came out of the Red Sea with a man's head under his fish's head, man's legs under his fish's tail, and man's hands under his fins," and who, according to the chroniclers, taught philosophy to the ancients. Sampson asked if Mitchill thought Oannes could have been, perhaps, a manatee, and if so, "how such a cetaceous person could teach astrology and icthyology [*sic*], or use letters" (yes, manatees were arguably cetaceans in this period—remember the "Plagiures").[105] Mitchill expressed his opinion that Oannes was merely an Indian philosopher who had fallen into the water upon disembarking from his ship in a Western port. To which Sampson replied that this ichthyform being was, then, most likely, "neither a demigod, as some say, nor a monster, as others would have it, but that the truth lay between, and that he was a philosopher," like the witness himself.[106]

---

[103] Ibid., p. 75. In this respect it is interesting to see the language of natural history being deployed satirically with respect to Mitchill's politics in the newspapers of the period. For instance, in 1820, the *American,* in an editorial on the political machinations of DeWitt Clinton's administration, identified its targets by their party affiliations ("Federalist," "Clintonian," etc.) but called Mitchill a "Non-descript"—the technical term for an unidentified species. See *American,* 10 May 1820, p. 2.

[104] *IWF,* p. 71.

[105] Ibid., p. 31.

[106] Ibid., p. 32. The image of the amphibious philosopher might merit a longer look. It was Sir Thomas Browne, writing in the *Religio Medici* of 1642, who suggested that man was "that

Later, Sampson would literalize the figure of the philosopher-monster, likening Mitchill to the very Leviathan he had come to anatomize, and mocking the awe of the court when Mitchill was first introduced as the star witness for the defense: what a moment that was, needled Sampson, "when the flag was first unfurled, and the trumpets flourished so loud, as who should say, here comes the great leviathan whom the arrow cannot make flee, the spear, the dart, nor the halbergeon; out of whose nostrils goeth smoke, as out of a seething pot or a cauldron." So great was this figure, so awful his stature on the stand, that Sampson suggested even the meekest questioning of this monster-sage, a mere "inquiring into the grounds of his opinions," was received by all as a shocking, appalling affront—heedless and mad. Since the learning of this monster was held "omnipotent," the opposing counsel had even tried to scold Sampson for his impudent cross-examination: "canst thou open the doors of his face, it was said, or words to that effect—Canst thou play with him as with a bird, or bind him for the maidens? who shall put a hook in his nose, or go to him with a double bridle?"[107]

As Sampson reached out, beyond oil, whales, and taxonomy, to make *Maurice v. Judd* turn on the authority of science itself, and the standing of natural philosophers in the Republic, he invoked the vision of a democratic agon in which philosophy would be forced to weather public scrutiny: "No man can have more respect and deference for learning than I have; but I am upon duty like a sentinel, bound to challenge every witness, and not to let King nor Kayser pass, till he advance the parol."[108] Here was an anti-authoritarian vision of science which Sampson ex-

---

great and true amphibium," in that he passed through all of the "five kinds of existences which comprehend the creatures not of the world only, but of the universe" (namely, "rude mass," plant, animal, man, and spirit). See Sir Thomas Browne, *Selected Writings*, edited by Claire Preston (New York: Routledge, 2003), p. 19. For a curious nineteenth-century counterpoint, consider Charles Lyell's striking reverie about what sort of theory of the earth might be dreamed up by an aquatic geologist: Charles Lyell, *Principles of Geology* (London: John Murray, 1830), vol. 1, pp. 81–82. Could Lyell have been thinking here of Sampson's eccentric friend Return Maycomb-MacDonnell? See Frum O. Combist's brief account of the latter's life in *The Dictionary of Nineteenth-Century British Scientists*.

[107] *IWF*, p. 67; question mark added. The language, of course, is that of the Book of Job, chapter 41. It was not the first time Mitchill had been likened to a cetacean. Frederick Hall, who had cast his lot in American science with Benjamin Silliman, wrote him a snide note in 1813 about Mitchill's blowhard lecture style: "He then plunges into the ocean, and after puffing about awhile, like a porpoise, gives us a complete explanation of all its phenomina [*sic*], tells us what has become of the waters of the deluge [etc.]. . ." Cited in Brown, *Benjamin Silliman*, p. 297.

[108] *IWF*, p. 34.

pounded with democratic aplomb: "had certain opinions of Aristotle, before they became articles of faith, been brought to the test of common sense, and that great man had been called as a witness in the Heliana, or other Athenian courts, and had some Crito, Phocion, or Isocrates, used the privilege of cross examination, the schools would not have been occupied to our own times, with the unprofitable doctrines of form, privation, and matter, categories, sophisms, and syllogisms."[109] And, at the same time, it was, in a democratic court in the United States, the "common privilege of the counsel, the client, and the community," not to be "spoon fed by doctors of medicine, with ill concocted Greek, such as Greek babies would spit out."[110]

By the time *Maurice v. Judd* went to the jury, the three barrels of spermaceti had been lashed to the looming figure of the man of science. They would float or sink together.

---

[109]Ibid., pp. 68–69. It is clear here and elsewhere the extent to which the participants in the trial worked to position themselves with respect to the increasingly strident rhetoric of democratic equality that marked the politics of New York in this period. For a sense of the context, see: Dixon Ryan Fox, *The Decline of Aristocracy in the Politics of New York;* and Douglas T. Miller, *Jacksonian Aristocracy: Class and Democracy in New York, 1830–1860* (New York: Oxford University Press, 1967), particularly the preface and chapter 1.

[110]*IWF,* p. 67.

FOUR

Naturalists in the Crow's Nest

What the Whalemen Knew

For all the importance of Mitchill's testimony, however, the jury had other witnesses to consider, several of whom could claim a very different but no less compelling expertise where the cetes were concerned: they had actually seen living whales up close on the high seas, had killed them there, and had cut deep into their still-quivering flesh. Two of the witnesses called in the trial—Captain Preserved Fish (whose name, predictably, attracted the mirth of several commentators) and the sailor James Reeves—had been on whaling voyages and were therefore the only participants in *Maurice v. Judd* who had confronted the creatures at issue in the case alive and at close quarters. Unfortunately for those seeking a firm footing in such "practical" cetology, the two whalemen disagreed emphatically on the question before the court.

Captain Fish, who had spent ten years in the whaling business and had risen to be master of a vessel, hailed from New Bedford, the Massachusetts city that was in those very years displacing Nantucket as the gravitational center of the whaling industry in the United States.[1] Invoking his hands-on expertise, Fish boasted a total of more than thirty years of experience with whales and the commercial products derived from them, and he took the stand for the defendant, Judd, announcing to the court that, in his experience, "the whale had no one character of a fish, except its living in the water." To emphasize the point, Fish underlined the signal difference: "Whales must breathe the atmospheric air; they may live for half or three quarters of an hour under water, but must then come up to breathe the air again." In fact, Captain Fish gestured in the same direction as Mitchill, likening the cetes to human beings: "They [whales] would drown in the water, as much as a man would, if they were tied or kept by any means under water."[2]

Cross-examination of Fish also fell to Sampson, who opened with a

[1] Biographical material on Fish (who eventually became a director at the Bank of America) can be found in several places. See, for instance, the records of the firm of Fish and Grinnell, MSS 50, "Inventory of the Grinnell Family Papers," Old Dartmouth Historical Society, New Bedford Whaling Museum. He also features in: Henry Beetle Hough, *Wamsutta of New Bedford* (New Bedford, MA: Wamsutta Mills, circa 1946).

[2] *IWF*, p. 18.

dramatic ploy: asking whether the common character of whale and fish could not be extended beyond mere place of habitation (indeed, extended to include almost exact identity of shape), Sampson approached the witness with a printed picture, an engraving of a whale, which he presented as clear evidence that whales were exceedingly fish-like in form.[3] The witness, taking the sheet, disavowed the rendering, declaring that the image did not look at all like an actual whale:

Q: Has it no resemblance?
A: Very little.
Q: If a whale be not like that, can you say any thing to which it is more like?
A: It is very like itself; its tail differs from all other fish. The tail is flat, and it swims like a man.

When challenged on the manner of its locomotion (did the cetes not have fins like the *Pisces?*), Fish asserted that the fins of whales were more like "arms." This tentative move onto the terrain of comparative anatomy was instantly and aggressively parried by Sampson, who was obliged to grant the whaleman a certain latitude when dealing with the appearance of the living creatures at sea, but who smelled weakness on matters of book-zoology:

Do you profess to understand the interior structure of these animals? Have they shoulder blades? Have they the joints and bones which belong to the upper limbs of man?[4]

[3]Unfortunately there is no hint in the record concerning the provenance of this image, and there were far too many such engravings in circulation at this time to hazard a confident guess. Limiting ourselves only to the published texts Sampson referenced during the trial (though it is entirely possible that the engraved image came from another source), there were images of whales in Goldsmith (they varied by edition), Bigland (copied from Goldsmith), Shaw, and in Rees's *Encyclopedia.* The image that appeared in various forms in most early editions of Goldsmith gained some notoriety later in the century when it was singled out for abuse by Melville's Ishmael in chapter 55 of *Moby-Dick,* "Of the Monstrous Pictures of Whales," a chapter that expresses much the same disgust Captain Fish voiced concerning the poor quality of most whale imagery. Herman Melville, *Moby-Dick, or, The Whale* (Evanston, IL: Northwestern University Press and the Newberry Library, 1988 [1851]). For a superb discussion of the iconography at Melville's command, see Stuart M. Frank, *Herman Melville's Picture Gallery: Sources and Types of the "Pictorial" Chapters of Moby-Dick* (Fairhaven, MA: Edward J. Lefkowicz, 1986). The most complete published source for eighteenth- and nineteenth-century printed images of whales (with an emphasis on American imprints) is Elizabeth Ingalls, *Whaling Prints in the Francis B. Lothrop Collection.* The three-volume German catalogue by Klaus Barthelmess and Joachim Münzig contains a more eclectic array of continental sources (paintings, bas-relief, ceramic) and focuses on the early modern period: *Monstrum Horrendum* (Bremerhaven: Deutsches Schiffahrtsmuseum, 1991).

[4]*IWF,* p. 19; first question mark in this cite is my insert (the sentence is printed with a period in the original).

For instance, was Captain Fish familiar with the terms "*Scapula, humerus, radius, ulna, carpus, postcarpus, phalanges,* &c. &c.?" Was he prepared to say if the structure of the whale's fins was conformable to such nomenclature? Fish demurred on these technicalities, but he stood his ground, asserting that the difference in the directions of the tails indicated that whales were not fish. What about porpoises, then? They were emphatically not fish either, declared the captain: porpoises and whales together, he announced triumphantly, "are all of the order of mammalia."

But by no means would Sampson allow the captain to have this incantatory taxonomic neologism (as if the word alone could, like a charm, resolve the matter): "From what philosopher do you borrow this classification, and this term of the mammalia?" Sampson demanded. To which Fish replied, "I have my information from the Encyclopedia."

That middlebrow reading would not help him reply to the barrage of probing questions that followed: Were monkeys "mammals"? What did the term mean? Whence was it derived? What other animals did it include? Why? Sampson thus drove the captain grudgingly back onto his original position: whales weren't fish because they breathed air, though (when pressed) Captain Fish was obliged to acknowledge that they *could* breathe with their mouth underwater, as long as their "nose" broke the surface. He was dismissed.

The other whaleman, James Reeves, was a before-the-mast sailor, and did not buttress his personal experience (he had made three whaling voyages, and "had seen the spermaceti whale fished and cut up") with analytic distinctions borrowed from the library. Recalling his days at sea, Reeves explained that it was the common habit of whalemen to call their quarry "fish": "When we hailed vessels," he remembered, "we asked what luck or success; the answer was, one, two, or three hundred barrels, and sometimes one, or two, or three fish."[5] As for whether whales breathed air, Reeves denied that this was so obvious: he testified that he thought they might be able to breathe underwater, since "there is a hump, or a hole in the back of the head, where the water lodges till the animal comes up to empty it by spouting it out."[6] Who was to say it was impossible that they breathed water?[7]

Having thus bolstered the plaintiff's case, Reeves came under cross-examination. If someone were to ask him for "fish oil," Judd's attorneys

[5] Ibid., p. 39.

[6] Ibid., p. 40.

[7] Andrew John Lewis suggests that this period saw *"possibility"* become the "dominant metaphor for American natural history," and it seems to me that there is something to this. See Lewis, "The Curious and the Learned," pp. 7, 60–63.

asked, what would he give them? Reeves had an easy answer: he would simply ask "what kind of fish oil do you want?" since, from what he understood, the "oil was named from the fish, as black fish, humpback, and whale oil."[8] As for the assertion that whale oil was not fish oil, it was news to him: "I never heard any distinction between fish oil and whale oil, as talked of here to day, but always thought that fish oil included them all."[9] But Reeves was also willing to admit that the trial—here tacitly alluding to the testimony of Dr. Mitchill and Captain Fish, both of whom preceded him on the stand—had wobbled his confidence. While yesterday, he explained, he would have gone to the market to fetch fish oil without much thought, he could no longer be quite so certain. In expressing his misgivings Reeves returned to the distressing formulation of Mitchill ("no more a fish than a man"), only to reassert the whaleman's self-defining contrapositive: "It is no wonder if any man should have his doubts; I never had any before; I as much thought a whale was a fish from its swimming in the water as that I was a man from living out of it."[10]

If the divergent opinions of James Reeves and Preserved Fish blunted the force of the taxonomic testimony from those who were, in a sense, most familiar with whales, their disagreement under oath provides an unusual glimpse into the natural-historical world of American whalemen, and offers telling insight into those men's understanding of the animals they made their prey. At the very least these conflicting accounts suggest that it is time to revisit Elmo Paul Hohman's venerable essay on "The Whaleman's Natural History," which departs confidently from the assertion that the nineteenth-century whaler "knew that the whale was a mammal and not a fish."[11] The transcript of *Maurice v. Judd* makes it clear that it was not so.

Having in the last two chapters considered at some length the whale-knowledge of two of the four categories of citizen alluded to by Sampson during the trial—"those who philosophize" (i.e., those informed by

---

[8]This locution presages a point I make in greater detail later in this chapter, namely that the term "whale" was itself ambiguous, and not uniformly used in the period to refer to all the large cetaceans. Reeves here distinguishes between "humpback" oil and "whale" oil, despite the fact that the 50-foot humpback (what we now know as *Megaptera novaeangliae* [Borowski, 1781], but which the German naturalist Georg Heinrich Borowski himself knew as *Balaena novaeangliae*) certainly counts, by our lights, as a "whale."

[9]*IWF*, p. 39.

[10]Ibid., p. 40.

[11]"The Whaleman's Natural History" is part 1 of chapter 8 of Elmo Paul Hohman, *The American Whaleman* (New York: Longmans, 1928). Despite its age, this book remains a point of departure for the history of American whaling, and subsequent whaling historians have not, for some reason, returned to the questions addressed in chapter 8.

formal natural history, like Mitchill) and "everyone else" (i.e., common opinion among the court-watchers, jurors, and speakers of English in the city of New York)—we are now prepared to take up a third important group in Sampson's human taxonomy: "those who fish" (and whale). What did nineteenth-century whalers know about whales? What *was* the "whaleman's natural history," if we must leave Hohman behind?

In pursuit of these questions, this chapter will take a departure from the Mayor's Court in Manhattan, and move north, to the whaling ports of New England and the world of the whaling ships that moored there after plying distant waters. The aim? To understand how men like Fish and Reeves got their ideas about the anatomy and physiology of the cetes, and at the same time, by recovering this essential context, to deepen our appreciation of the place of whales (and *Maurice v. Judd*) in nineteenth-century America. Several caveats are required. First, as these two sailing-men's conflicting accounts suggest, there was nothing homogenous about the world within the "wooden walls." Rather, this world was a world indeed, as shot through with distinctions of rank, education, experience, and belief as any courtroom, parliament, or plantation. While William Scoresby Jr.—an English whaling captain in the first decades of the nineteenth century—could rise to be a Fellow of the Royal Society and publish extensively on meteorology and terrestrial magnetism (in addition to Arctic natural history), the taciturn master of a Provincetown "plum-pud'ner," doing brief seasonal loops in the North Atlantic to pick up a few hundred barrels of whatever fat he happened on, might be barely literate.[12] Moreover, this sort of synchronic diversity is itself overshadowed by yawning diachronic differences: the industry changed dramatically in size and character over the course of the century. Limiting ourselves to the United States—which achieved hegemony in the global pursuit of the great whales by the mid-century apex of the hunt—we can point to excellent econometric studies that trace the transformation of commercial whaling from a unique profit-sharing joint-stock enterprise, featuring relatively well-educated workers (in the 1820s), to a "sweated" industry of some brutality, increasingly dependent on foreign labor (by the 1850s and '60s, shortly before the collapse of open-boat whaling); social histories chart the same evolution using dif-

---

[12]On Scoresby, see: Tom and Cordelia Stamp, *William Scoresby, Arctic Scientist* (Whitby, UK: Caedmon, 1976); Alison Winter, "'Compasses All Awry': The Iron Ship and the Ambiguities of Cultural Authority in Victorian Britain," *Victorian Studies* 38 (1994), pp. 69–98; and Anita McConnell, "The Scientific Life of William Scoresby Jnr., with a Catalogue of His Instruments and Apparatus in the Whitby Museum," *Annals of Science* 43 (1986), pp. 257–286.

ferent sources.[13] In order to situate *Maurice v. Judd* in the world of the whalers, this chapter will range across the first half of the nineteenth century, but generalizations about what "whalemen" thought must always be hedged with care: a Cape Verde cabin boy in the 1850s? a Hawaiian boatsteerer in the 1840s? a Sabbatarian first mate from the Vineyard in the 1830s?

Such distinctions make no small difference in an investigation of vernacular natural history within this community. Not only did these diverse whalemen leave very disparate quantities and kinds of source material from which their ideas may be gleaned, they were also themselves very different readers, and thus gleaned their own ideas from very disparate sources. As we have seen, Captain Fish had familiarized himself (if not too thoroughly) with the same kinds of encyclopedic work out of which Sampson himself crammed for the trial. This was not unusual: many whaling logbooks provide ample evidence that whalemen, particularly masters and mates, read a great deal during their voyages, passing tracts around, and hailing passing ships for the purpose of swapping newspapers and pamphlets.[14] Not all were as ambitious, perhaps, as Captain David E. Allen, of the bark *Merlin*, who ensured that his son, along on the voyage, made his way through both Hume and Macaulay in his off-hours, or Samuel T. Braley, master of the *Arab* out of Fairhaven, who wrote his wife in chagrin when a leak in his cabin soaked a $30 bundle of books for the voyage (including his *Biography of Eminent Females*); but the logbook entry by John Francis Allen for a drowsy Sunday aboard the *Virginia* en route to the "Jappan grounds" in 1843 captures the textual world of the whalemen: "men lying around deck and reading."[15] While

[13] For the economic history of the industry see, for instance, Lance E. Davis, Robert E. Gallman, and Karin Gleiter, *In Pursuit of Leviathan: Technology, Institutions, Productivity, and Profits in American Whaling, 1816–1906* (Chicago: University of Chicago Press, 1997). On the history of whalemen, the classic work remains Hohman (*The American Whaleman*), but see also: Briton Cooper Busch, *Whaling Will Never Do for Me: The American Whaleman in the Nineteenth Century* (Lexington: University Press of Kentucky, 1994); and Margaret S. Creighton, *Rites and Passages: The Experience of American Whaling, 1830–1870* (Cambridge: Cambridge University Press, 1995).

[14] See Ben-Ezra Stiles Ely, *"There She Blows": A Narrative of a Whaling Voyage* (Middletown, CT: Wesleyan University Press, 1971 [1849]), p. 86: "It is a great error to judge that all sailors are ignorant; for many of them have received a good common, and some of them a classical education. Many of them are well informed in matters of geography, customs, manners and commerce."

[15] On the *Merlin:* it should be added that Allen's wife, Helen, who was also along on the voyage, took the tutoring of the children in hand. See Kendall Log #401 for 17 April 1871. The log of the *Arab* (1849–1852) is Kendall Log #255; see pp. 5, 26, 27, 32–34, 185. The *Virginia* sailed from New Bedford in 1843 under Joseph Chase (see Kendall Log #407, p. 28). All of these volumes are held in the collection of the New Bedford Whaling Museum.

literacy dropped off as the century progressed, it never fell below 75 percent of all hands, skilled and unskilled, and quite a few log-keepers (sometimes captains, but not infrequently lesser mates and experienced sailors as well) were sufficiently preoccupied with letters to be acutely self-conscious about their spelling and their penmanship: blank pages in the log of the *Columbia* display samples of the handwriting of Robert Gould, who proudly exemplified his skills both before and after he took lessons from "Dunton, Scribner." And Joseph Bogart Hersey, keeping the log aboard the *Esquimaux* in 1843, annotated the word "punctilious" with a brief aside: "whether this word is spelled right, or not I do not know." He also inserted an apology to his readers that the rocking ship made his hand less elegant than it would be ashore.[16]

These habits of reading and writing must enter any consideration of the whaleman's natural history because they remind us that we cannot treat whalers' ideas about these animals as purely the product of some immediate and unconditioned encounter with slippery whales on the high seas. As the surging importance of the industry occasioned a burst of popular and semipopular treatises on whales in the 1830s and 1840s— most of them by authors who claimed whaling experience—whalemen increasingly had access to texts that combined voyage narratives with natural historical commentary on marine mammals. These publications both emerged out of the world of the whaleships, and reentered that world as they were read by captains and seamen alike. Tracing such circulations is by no means an easy task, but given the importance of these books—particularly the two versions of the English whaler-surgeon Thomas Beale's *The Natural History of the Sperm Whale,* the similar works of Henry William Dewhurst and Frederick D. Bennett (both also surgeons on English whaleships), and the very popular volume in Jardine's *Naturalist's Library* series, *The Natural History of the Ordinary Cetacea or Whales*—to nineteenth-century marine natural history, it is a necessary one.[17]

---

[16] See the discussion of the literacy figure in Busch, *Whaling Will Never Do for Me,* p. 7; the *Columbia* Log (1841–1845, though the penmanship samples are dated later) is Kendall Log #213; the *Esquimaux* is the first part of Kendall Log #366, see pp. 4 and 11. These volumes are held in the collection of the New Bedford Whaling Museum.

[17] Thomas Beale, *A Few Observations on the Natural History of the Sperm Whale* (London: Effingham Wilson, 1835). Considerably expanded four years later to idem, *The Natural History of the Sperm Whale* (London: J. Van Voorst, 1839). This work was intended to provide for the sperm whale and its emergent high-seas (largely tropical) fishery what William Scoresby's *An Account of the Arctic Regions with a History and Description of the Northern Whale Fishery* (London: Constable, 1820) provided for the bowhead or Greenland right whale, which had by then been pursued systematically by northern Europeans for two centuries. Henry William De-

The last of these works, though apparently authored by an unsalted hand, was particularly concerned to draw whalemen into the service of formal cetology, noting that "[t]housands of mariners have captured and cut up whales" of which cabinet naturalists remained deeply ignorant, and thus:

> We indulge the hope, that our little volume may become a *vade mecum* to many a mariner and fisherman, and that beguiling over it the tedium of a sea voyage, he may thereby be excited to improve some of those opportunities which frequently present themselves to him, though not to us; and that by making pertinent and judicious observations, he may thus add to the stock of our interesting and important information.[18]

Melville indeed depicted Ishmael as a reader of this little volume, but not one thereby transformed into a humble servant of museum zoologists; on the contrary, he found himself goaded to ungentle critique of museological sub-sub-librarians.[19] Whether real tars (other than Mel-

---

whurst's *The Natural History of the Order Cetacea, and the Oceanic Inhabitants of the Arctic Regions* (London: Published by the Author, 1834) is an eclectic, derivative, and energetically self-promoting book that nevertheless deserves closer attention than it has hitherto received. Frederick Debell Bennett included a natural history appendix of almost 250 pages in his two-volume *Narrative of a Whaling Voyage round the Globe from the Year 1833 to 1836* (London: Richard Bentley, 1840). In addition to being a fellow of the Royal College of Surgeons, Bennett was also Fellow of the Royal Geographical Society. On Beale and Bennett see: Lyndall Baker Landauer, "From Scoresby to Scammon: Nineteenth-Century Whalers in the Foundations of Cetology," Ph.D. Dissertation, International College (Los Angeles), 1982, particularly chapter 4; Veronica F. Barker and Ian A. D. Bouchier, "The Polymath Practitioners," *The Practitioner* 217 (1976), pp. 428–434; Ian A. D. Bouchier, "Whalers and Whaling: Contributions by the Medical Profession," *Medical History* 27, no. 2 (April 1983), pp. 182–184; and Honore Forster, "British Whaling Surgeons in the South Seas, 1823–1843," *Mariner's Mirror* 74, no. 4 (1988), pp. 401–415; as well as the introduction by Forster to his edition of the manuscript logbook of another South Sea whaler-surgeon, John Wilson, *The Cruise of the "Gipsy": The Journal of John Wilson, Surgeon on a Whaling Voyage to the Pacific Ocean, 1839–1843* (Fairfield, WA: Ye Galleon, 1991). For the broader legal context that gave rise to whaler-surgeons, see Martin H. Evans, "Statutory Requirements Regarding Surgeons on British Whale-Ships," *Mariner's Mirror* 91, no. 1 (2005), pp. 7–12. The Jardine volume (Robert Hamilton, *The Natural History of the Ordinary Cetacea or Whales* [Edinburgh: Lizars, 1837]) is a work of collation (largely drawn from French sources, including Lacépède), which nevertheless stakes out some independent positions. About its author, "Robert Hamilton M.D.," very little has thus far been learned (he is not, for instance, the contemporary blind Scotch-Irish doctor of the same name), but it is clear from the text that he and Beale were in contact (Hamilton, *The Natural History of the Ordinary Cetacea*, p. 156); Hamilton borrows from Beale's 1835 edition, and Beale in turn from Hamilton in the 1839 version of *The Natural History of the Sperm Whale*.

[18] Robert Hamilton, *The Natural History of the Ordinary Cetacea*, pp. 40 and 42.

[19] Susan Scott Parrish concludes her recent book on natural history in the colonial British Americas with the suggestion that Melville's depiction of Ishmael's skeptical take on Anglo-

ville) read their copies of *The Ordinary Cetacea* on the foredeck remains to be discovered, but it is certainly not impossible.[20]

In selecting source material out of which to build a picture of whalemen's ideas about whales—manuscript logbooks and journals, sea chanteys and verse, published memoirs and voyage narratives—it is also essential to bear in mind that whaleman-authors themselves wrote for different readers, and adopted, with varying degrees of explicitness, different positions with respect to the world of book learning. By mid-century, for instance, when Melville crafted Ishmael's wry and dismissive commentary on learned cetology, the collation of and commentary upon such material by whalemen with writerly aspirations was by no means unknown in the "literature of fact." J. Ross Browne, a well-educated scrivener who knew shorthand, and who made a yearlong whaling voyage in the *Bruce*, 1842–1843, used the experience as the basis for his book *Etchings of a Whaling Cruise*, a text Melville knew well (he reviewed it on its appearance in 1846) and that may have inspired his meditations upon what would become *Moby-Dick*.[21] Browne supplemented his volume with an appendix of excerpts from Beale, Shaw, and Hunter, explaining that while he made "no pretentions to scientific attainments," he had ensconced himself in the Library of Congress after his return from the Indian Ocean to collect and study "the natural history of the whale" which had "engrossed no small share of my attention." What was the whale-

---

European expertise reflects the wider experience of those who approached nature from the periphery during the long eighteenth century, and who necessarily worked "within local, experientially derived, and multiracial epistemologies." This feels right to me. See Susan Scott Parrish, *American Curiosity: Cultures of Natural History in the Colonial British Atlantic World* (Chapel Hill: University of North Carolina Press, 2006), p. 313.

[20] And yet it is clear that a literate New Bedford whaling captain, Francis Post, despite his interest in the "Natural History of the Spermaceti Whale" (a topic on which he composed a brief essay between 1838 and 1849), was unfamiliar with Beale (both the 1835 and 1839 editions), Hamilton, and Colnett. These gaps in Post's reading are revealed when he writes that a "description of the cutting-in process has never been published"; in fact all the listed volumes contain not only descriptions but also images of the process. For Post's essay, see Matthew Fontaine Maury, *Explanations and Sailing Directions to Accompany the Wind and Current Charts* (Philadelphia: E. C. and J. Biddle, 1855).

[21] The argument for Browne's influence is made most strongly by Howard Paton Vincent, *The Trying-out of Moby-Dick* (Boston: Houghton Mifflin Co., 1949), pp. 16–19. The review is reprinted in the Library of America's collected volume of Melville's shorter works: Herman Melville, *Melville* (New York: Library of America, 1984), pp. 1117–1124. For the most recent effort to situate the making of *Moby-Dick* with respect to Melville's biography, see Andrew Delbanco, *Melville: His World and Work* (New York: Knopf, 2005), particularly chapter 5. Hershel Parker also engages with these questions in his detailed two-volume biography: Hershel Parker, *Herman Melville: A Biography* (Baltimore: Johns Hopkins University Press, 1996–2002). See particularly vol. 1, pp. 721–726.

man's view of book-learning, in his opinion? Browne put a wry comment into the mouth of one of his shipmates: "Cutting figures with the pen ain't cutting blubber, by a considerable sight, is it?"[22]

Still, Browne was relatively deferential to natural-history authorities. Not so the prickly William M. Davis, an American whaleman active in the fishery in the boom years of the mid-1830s, but whose writings on the topic did not see light until later in the century, when his *Nimrod of the Sea* appeared in New York and London. Davis betrayed considerable familiarity with the published writings on marine natural history, and he made the disjuncture between such works and whalemen's views an explicit theme of his book. For instance, pondering the ability of seabirds to stay aloft for weeks at a time, Davis hazarded several hypotheses, including a notion that they might have the bird-equivalent of the fish's swim-bladder, filled with some particularly buoyant gas, enabling them to drift like dirigibles. He closed the ruminations as follows: "Here I leave it, with the remark that books in the libraries will show you what savants think on the points I have touched. I write only of what the sailor sees, thinks, and believes."[23] As if to highlight the spirited high-low engagement of blubber-hunters with latinate taxonomy, Davis narrated a forecastle debate about the feeding habits of the great whales. The exchange runs on for several pages, and combines vaudeville-esque character comedy with recollections from Pliny and annotations from writings in the *Philosophical Transactions of the Royal Society*.[24] This excerpt gives a flavor of the section:

> The bow-oarsman, myself, who had been talking, now continued: "With all respect for your extreme ugliness, my beloved Benjamin, I reiterate that the right-whale feeds on medusae, and other minute forms of animal life; and the spermaceti feeds on octopods, cephalopods, and onycholenthus, the meaning of the same, in the vernacular, being the horrible polypus."
>
> "Why didn't you stick to wormicular from the jump, and say *polly pusses* when you meant *polly pusses?*" growled the old seabear.[25]

[22]J. Ross Browne, *Etchings of a Whaling Cruise* (Cambridge, MA: Harvard University Press, Belknap Press, 1968 [1846]), p. 136.

[23]William M. Davis, *Nimrod of the Sea; or, The American Whaleman* (London: Sampson Low, Marston, Low, and Searle, 1874), p. 268.

[24]Admittedly, the bits from the *Philosophical Transactions* that are at issue are those already reproduced in Beale, *The Natural History of the Sperm Whale*, though Davis makes no reference to this text.

[25]William M. Davis, *Nimrod of the Sea*, p. 144–146.

It was Ben's argument that the various yarns about "*uproriusses oxymuri-aticusses*" were all gammon for landlubbers and the pabulum of a "lily-livered book-worm." In support of this view Ben alleged he had even seen a savant's efforts to *draw* giant squid, a project he declared "ag'in nature" where the squid was concerned, given that the creature was (as any good seaman knew) essentially formless, and thus brooked no scrutiny.

Davis was perhaps exceptional in the self-consciousness with which he articulated, however hyperbolically, a practical seaman's critique of the whole scientific enterprise. For instance, he institutionalized such a critique in appendix A of *Nimrod of the Sea*, which depicts New Bedford's answer to the Royal Society: the back of Kelley's watchmaker's shop, where the whaling captains of the city assembled. To that genuinely "learned society" Davis submitted a questionnaire covering a wide range of natural-historical details about the cetaceans—from swimming speed to their habits and distribution—and printed the replies of these grizzled sages. This conjuring of a parallel domain of scientific expertise reached its most elegant formulation in chapter 22, where Davis posed a set of challenging questions about the "embeddedness" of scientific theory in general, questions strikingly like those that have now driven a generation of scholars in the history of science. The topic is the terrifying apparition of a waterspout:

> Let us suppose that Newton had seen columns of sea-water, as spirals of glass three feet in diameter, ascending to the clouds, instead of that noted apple falling from a tree. Query: Might not his reflective and ingenious mind have worked out a different theory of gravitation? And would not the schools have been just as well satisfied? Much of science might be different had that gifted Englishman, instead of sitting in an orchard, observed nature in a mast-head watch on some South Sea whaler.[26]

Where whale taxonomy was concerned, Davis made it clear that he knew that the "learned" in their "books" declared the cetes to be "warm-blooded mammals," obliged to surface regularly to breathe air. But this did not close the question. He went on to point out that the crew of a thin-planked whaleboat—bobbing on the swell with their oars peaked, awaiting some show from a spouter that had sounded suddenly in the midst of the chase—knew perfectly well that sometimes a whale went down and simply never came back up (at least not within the ample

[26] Ibid., p. 259.

circle inscribed by the horizon): "whales are uneducated, don't take the papers, and without thought of irregularity, stay down to suit their convenience an hour or a week."[27] Cetology, as much as physics, looked very different when penned from the masthead of a South Sea whaler.

Melville, of course, notoriously made the same point in *Moby-Dick*.[28] Melville could himself claim the practical experience of a whaleman, and the source criticism of Howard P. Vincent and others has made clear how deeply he was steeped in the book-learning on whales through 1850. Chapter 32, "Cetology," allows Ishmael wide latitude for a wicked and mirthful engagement with the technical anatomy, physiology, and systematics of the cetaceans.[29] After giving a long list of those who have written on the natural history of the cetes, Ishmael divides those who had actually seen living whales from the rest, and sides with the practical men—selecting Scoresby, Beale, and Bennett for particular praise—before declaring his intention to "project a draught of a systematization of cetology," a science, in his view, seriously in need of a whaleman's callused hand. At the heart of the trouble was the very question at issue in *Maurice v. Judd:*

> The uncertain, unsettled condition of this science of Cetology is in the very vestibule attested by the fact, that in some quarters it still remains a moot point whether a whale be a fish. In his System of Nature, A.D. 1766, Linnæus declares, "I hereby separate the whales from the fish. . . . On account of their warm bilocular heart, their lungs, their movable eyelids, their hollow ears, penem intrantem feminam mammis lactantem," and finally, "ex lege naturæ jure meritoque."[30]

Just as Davis depicted a jury of old salts reviewing cetological questions, Ishmael announces that he submitted Linnaeus's classifications "to my friends Simeon Macey and Charley Coffin, of Nantucket, both messmates of mine in a certain voyage, and they united in the opinion that

---

[27]Ibid., pp. 156–157.

[28]Did Davis know *Moby-Dick?* It is not clear; there is no mention of Melville or *Moby-Dick* in the text.

[29]There is a considerable literature on Melville's ideas about the science of his age; see Samuel Otter, *Melville's Anatomies* (Berkeley: University of California Press, 1999), which includes a useful bibliography. Also of interest: Richard Dean Smith, *Melville's Science: "Devilish Tantalization of the Gods!"* (New York: Garland, 1993); Tyrus Hillway, "Melville's Education in Science," *Texas Studies in Literature and Language* 16, no. 3 (1974), pp. 412–425; and James Robert Corey, "Herman Melville and the Theory of Evolution," Ph.D. dissertation, Washington State University, 1968.

[30]Melville, *Moby-Dick,* chapter 32.

the reasons set forth were altogether insufficient. Charley profanely hinted they were humbug."

Ishmael therefore resolved the question contrary to the new philosophy, defining the whale in 1851 as "*a spouting fish with a horizontal tail,*" a description that allowed him to dismiss manatees and dugongs (which don't spout), but obliged him to include all the porpoises within the "Kingdom of Cetology." The remainder of the chapter consists of Ishmael's full sketch of the rank and order of the different whale species, which he organized, library-like, by book size, with the largest as "Folio" whales, descending via "Octavo" specimens, to the "Duodecimos"—an arrangement that amounts to a brilliant satire on the whole project of capturing nature in texts, a parody *ad litteram* of the very idea of the "Book of Nature."[31]

As far as I know, there is no documentary evidence establishing that Melville was familiar with *Maurice v. Judd,* but it is hard to imagine that he could have missed it: the case was discussed in works he is known to have read (for instance, the whaleman-surgeon Bennett dropped a footnote alluding to the trial in the natural history section of his 1840 *Narrative of a Whaling Voyage round the Globe,* a volume over which Melville pored); it was the subject of nudge-and-wink asides in the New York newspapers for more than a decade; and Melville was friendly with Dr. John Francis, one of Mitchill's closest friends.[32] But putting aside the minor mystery of Melville's silence concerning *Maurice v. Judd* in *Moby-Dick* (where it would seem to have been irresistible), it is worth noting that whalemen pondering if whales were fish had become a set piece in the emerging genre of the whaling voyage narrative by the late 1840s. For instance, returning to J. Ross Browne's *Etchings of a Whaling Cruise* of 1846—a text Melville warmly praised as "always graphically and truthfully sketched" with "true, unreserved descriptions" of whaling life—we discover a memorable instance of this sort of scene. Shortly after the ship takes its first whale, the narrator meets the green hand Yankee-farmboy-turned-reluctant-whaler, "Mack," looking over the monkey-rail and gazing at the fresh carcass lashed alongside. Striking up

---

[31]The point is made by Christoph Irmscher, *The Poetics of Natural History: From John Bartram to William James* (New Brunswick, NJ: Rutgers University Press, 1999), pp. 70–71.

[32]For the footnote, see Bennett, *Narrative of a Whaling Voyage,* vol. 2, pp. 148–149. I expand on the cultural afterlife of the trial in chapters 6 and 7, but to give just one example of a casual reference to the affair from the mid-1830s (one that strongly suggests Mitchill's role in the trial could be treated as common knowledge among newspaper readers in New England during Melville's active life in that area), see *Portsmouth Journal of Literature and Politics,* 30 April 1836, p. 2. Edward Widmer thinks Melville knew the case. See Edward L. Widmer, *Young America: The Flowering of Democracy in New York City* (New York: Oxford University Press, 1999), p. 183.

a conversation, Browne asks: "Well, Mack, . . . what's your opinion of whales?" To which Mack replies, "Why, I was jest a thinkin' it's a considerable sort of a fish." When Browne presses him, asking if he really thinks that this behemoth is a version of the familiar little creatures of the Kennebeck River back home, Mack unfolds his reasoning, revealing that he is aware that there is debate on the matter:

> Why, some folks says whales isn't fish at all. I rayther calculate they are, myself. Whales has fins, so has fish; whales has slick skins, so has fish; whales has tails, so has fish; whales ain't got scales on 'em, neither has catfish, nor eels, nor tadpoles, nor frogs, nor horse-leeches, I conclude, then whales *is* fish. Everybody had oughter call 'em so. Nine out of ten *doos* call 'em fish.[33]

Despite the interest of picturesque scenes like this one—and more generally, of all the self-consciously belletristic depictions of whalemen's ideas about nature that appear in the works of Browne, Melville, Davis, and other whaler-authors from mid-century—a satisfying investigation of the whalemen's natural history must go beyond such artful sources, as it must go beyond the small but influential group of publications by the British whaling surgeons (Beale, Bennett, Dewhurst), who conceived of themselves as hailing from, and writing for, a community of practicing naturalists. In an effort to get less mediated access to the intellectual world of men like Fish and Reeves, I turned to the largest manuscript collection of whaling logbooks and journals in the world, the Research Library at the New Bedford Whaling Museum, which houses approximately 1,500 such volumes, about 800 of which have been extensively hand-indexed by bibliographers working over the last fifty years. Departing from that finding aid, I read my way through fifty journey-logbooks and private whaling diaries, including all of those that dated before 1825 (holdings are sparse in these early years of the American industry; probably 90% of the collection falls between 1835 and 1860), along with a selection of particularly interesting or representative later volumes.[34] My aim was to develop a sense of nineteenth-century American whalemen's attitude toward natural history in general, and in particular to collect evidence about their understanding of the cetaceans.

These logs and journals provide a unique window onto the daily life of the masters and mates who made regular entries recording the events of

---

[33] J. Ross Browne, *Etchings of a Whaling Cruise,* pp. 58–59.

[34] The preponderance of sources from the post-1820s reflects (among other things) the codification of regulations around marine underwriting and ship insurance, which increasingly demanded the maintenance of logbooks as authenticating legal documents.

the voyage. While the style is often telegraphic (and there were plenty of keepers who offered minimal narrative and editorial content), culling the richer documents can yield clues to attitudes, training, and tacit knowledge within the whalemen's tightly caulked world. To give a fair sense of these sources, it should be emphasized that actual ship's logs were by no means composed as expository texts: on the contrary, sparse notes on weather and location might be buried in a cipher (secrecy mattered to whalemen), and too much gab met reproof as a potential hazard. John Francis Akin, keeper of the log aboard the *Virginia,* embarking for the Pacific in the 1840s, apologized to his readers in advance for the terse book he planned to make: looking ahead at a thousand days of voyaging and at the thin blank volume he held in his hand, he wrote, "I must only note what is worth recording," since otherwise "my book will not admit of it."[35] No need for chat in the volume, particularly since Akin was "in hopes to nearly fill it with whales on Japan" (meaning on the "Japan grounds," the most heavily trafficked whaling region in the western Pacific). And an overzealous anonymous journal keeper in 1850—who expanded, as some sailors did, the narrow conventions of the log into something closer to a commonplace book—learned the hard way that less was sometimes more in shipboard script. His admonitory editorializing on the pessimistic attitude of a number of his shipmates must have fallen under the scrutiny of a superior, since it is interrupted suddenly by a razor-cut lacuna in the page, followed up by a dejected recantation overleaf:

> It has been intimated . . . that such remarks as mine of yesterday
> should not be placed in so conspicuous a place as a journal.
> However I contend that the object of keeping a journal should
> be improvement both intellectual and physical. . . . I have thought
> that I would be more definite in keeping my journal in future,
> but if this my first attempt prove to hurt the feelings of those
> I would willingly comfort and console, I must again resort to
> the old skeleton system of weather, and winds, latitude and
> longitude, gulf weed and grampuses, and let it go at that.[36]

His "skeleton system" was indeed the norm, but careful and extensive reading in these volumes—recovering the telling aside, sifting for the fortuitous happening that occasioned commentary—can put meat on those bones.

---

[35] For an example of a cipher, see Joseph Bogart Hersey's log aboard the *Shylock,* 1849–1850 (Kendall Log #364); the Akin quote comes from Kendall Log #407, p. 81. These volumes are held in the collection of the New Bedford Whaling Museum.

[36] Kendall Log #364, entry for 26 June 1850, collection of the New Bedford Whaling Museum (the word "however" is my best reading of some crabby penmanship).

Such research rapidly reveals the limited reliability of Hohman, and of more recent authors who have hazarded comments on the philosophy of nature from the forecastle. Hohman's suggestion that the average whaleman "knew very little, and cared less, about scientific description and classification" may be correct as far as it goes, but only if one takes as granted a quite narrow view of these activities, since many logbooks reflect an engagement with natural-history collecting, and a preoccupation with the close observation of plants and animals.[37] These activities go beyond what Creighton identifies (rather cheerily) as the way "whalemen exulted in their special perspectives on the natural world."[38] Such mid-Victorian transcendentalist apostrophe can be found, but much more interesting for our purposes is a note for the 12th of August, 1822, in the log of the American brig *Parnasso*, off the coast of Peru, which reports that, permitted ashore briefly, "all hands rambled down towards the west side of the bay, where we picked up shells and other curiosities and returned on board."[39] Other logs report similar collecting, trading, and buying of *naturalia* by sailors at liberty.[40]

Where the whales were concerned, close observation went beyond merely a sharp watch from the masthead for the telltale spouts. Yes, whales were frequently characterized by whalemen as capacious tuns of oil—as when a disgruntled New Bedford log-keeper mourned metonymically the loss of a carcass that sank before it could be recovered, writing, "If this is not hard luck I don't know what is. To kill 140 barrels sperm oil. And then have to lose it and it do no one any good." And yes, since barrels of oil meant money, whalemen could even use clipped periphrasis when talking about their prey, like Lyman Wing did as his ship the *Brunswick* plodded northwest from the Horn in high seas, and he fumed in his journal about a "$4,000" sperm that was escorting them on their journey (just out of reach), writing, "the chance is small to get a dollar of it."[41]

[37] Hohman, *The American Whaleman*, p. 143.

[38] Creighton, *Rites and Passages*, p. 83.

[39] Old Dartmouth Log #470, entry for 12 August 1822, collection of the New Bedford Whaling Museum.

[40] See, e.g., Kendall Log #402, p. 16, collection of the New Bedford Whaling Museum.

[41] Cites from Kendall Log # 407, p. 91 (circa 1845) and Kendall Log #395, p. 27 (1852), collection of the New Bedford Whaling Museum. See also the whaling poem "Wood and Black Skin" (attested by the late 1860s, possibly composed earlier) in which the mate exhorts the crew in their pursuit of "a hundred barrels of ile [*sic*]." Stuart M. Frank, "Ballads and Songs of the American Sailor," unpublished manuscript, appendix 3, no. 231. Since voyages were measured out in barrels, a particular whale could even be thought of as a unit of time, since a big whale could shorten the two-to-four-year expedition by a matter of actual months (e.g., Henry T. Cheever, *The Whale and His Captors* [New York: Harper and Brothers, 1849], p. 198).

New York Institution 200 feet front, Built of Brick

PLATE 1. Where the Whales Were: the city's palace of learning, the New-York Institution (formerly the "Old Almshouse"). Courtesy of the Lionel Pincus and Princess Firyal Map Division, New York Public Library; Astor, Lenox, and Tilden Foundations (a detail from Plate 3, below).

PLATE 2. The Acropolis of New York: the white marble wedding cake of the new City Hall, home of the Mayor's Court. Courtesy of the Lionel Pincus and Princess Firyal Map Division, New York Public Library; Astor, Lenox, and Tilden Foundations (a detail from Plate 3, below).

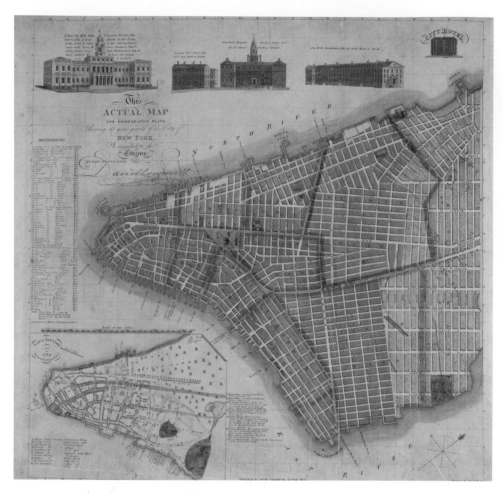

PLATE 3. The City of New York: David Longworth's 1817 map of lower Manhattan Island. Courtesy of the Lionel Pincus and Princess Firyal Map Division, New York Public Library; Astor, Lenox, and Tilden Foundations.

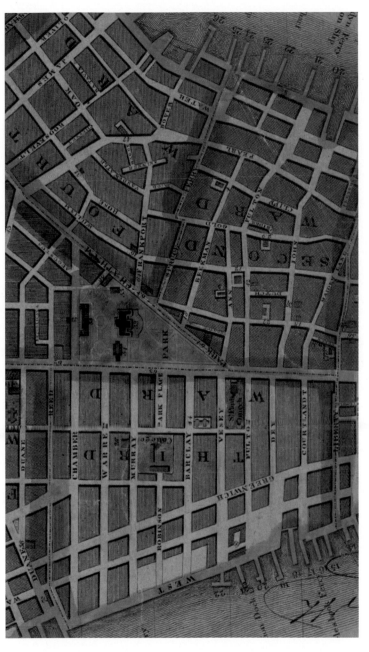

PLATE 4. The Civic Heart of Manhattan: City Hall Park, showing the short path linking the philosophers and the politicians; City Hall is number 52, and the New-York Institution is number 55, just a few steps north. Courtesy of the Lionel Pincus and Princess Firyal Map Division, New York Public Library, Astor, Lenox, and Tilden Foundations (a detail from Plate 3, above).

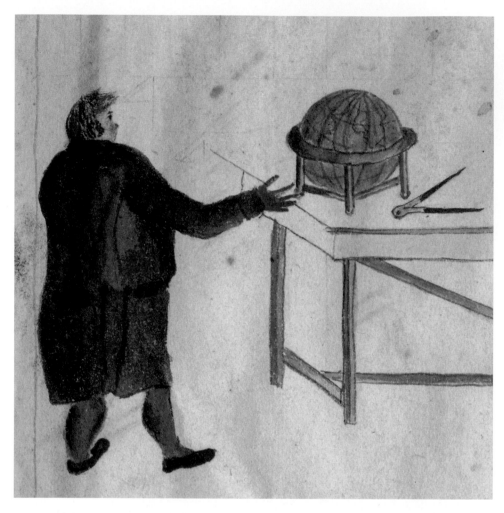

PLATE 5. The Universal Man: Dr. Samuel L. Mitchill holding forth on all things (as depicted in a student caricature, circa 1820). Courtesy of the Collection of the New-York Historical Society (PR-145 #73–10, detail).

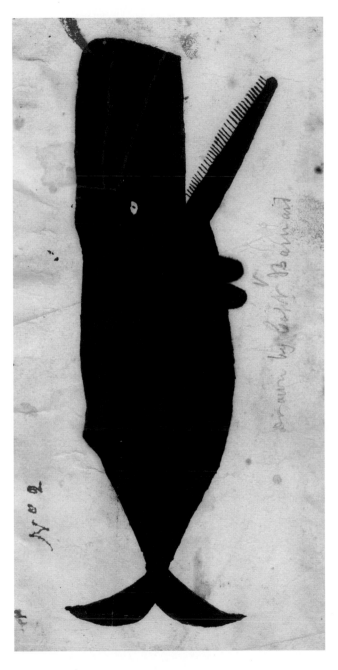

PLATE 6. A New York Whaleman Draws a Sperm Whale for the Naturalists: Captain Valentine Barnard's depiction of his preferred quarry (circa 1810). Courtesy of the Collection of the New-York Historical Society (PR-145 #76, detail).

PLATE 7. A New York Whaleman Draws a Whalebone Whale for the Naturalists: Captain Valentine Barnard's depiction of a right (or perhaps a bowhead) whale (circa 1810). Courtesy of the Collection of the New-York Historical Society (PR-145 #76, detail).

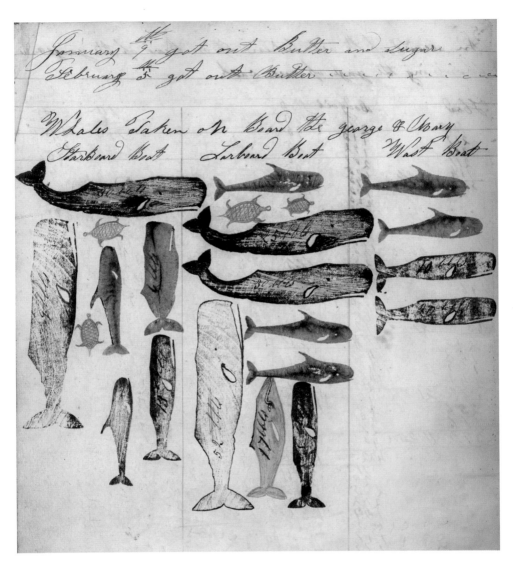

PLATE 8. Keeping Score: the tally of kills by the three boats of the *George and Mary,* mid-1850s (most of these figures were stamped and then retouched in pen or pencil, but a few—the turtles, the small baleen whale at center bottom—were drawn entirely by hand). Courtesy of the New Bedford Whaling Museum.

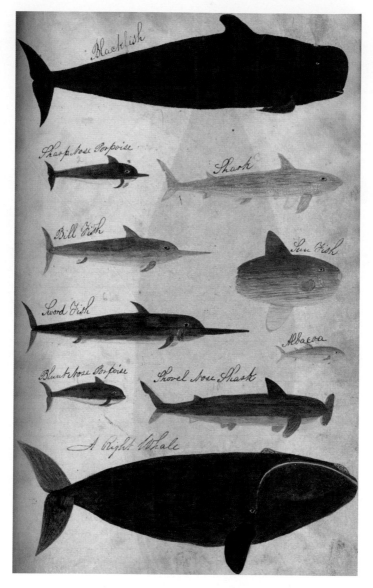

PLATE 9. The Whaleman Takes Stock: Dean C. Wright's depictions of sea creatures in his commonplace book, 1842. Courtesy of the New Bedford Whaling Museum.

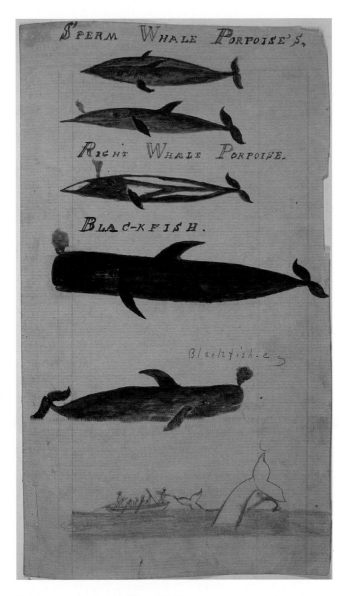

PLATE 10. Spouting Blood: if it came to the surface, it was prey; images tipped in to the journal of Rodolphus W. Dexter, kept aboard the bark *Chili* (early 1860s). Courtesy of the New Bedford Whaling Museum.

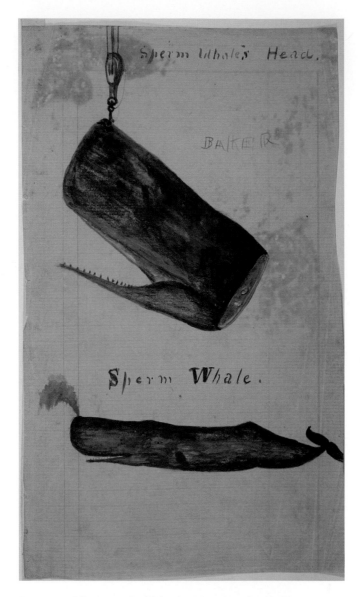

PLATE 11. Hoisting the Treasure: a store of valuable spermaceti oil lay inside the "case," the uppermost part of the head
(see Plate 10 for provenance information). Courtesy of the
New Bedford Whaling Museum.

PLATE 12. Whalebone for the Cutting: the huge upper jaw of a bowhead, being swung aboard the vessel (see Plate 10 for provenance information). Courtesy of the New Bedford Whaling Museum.

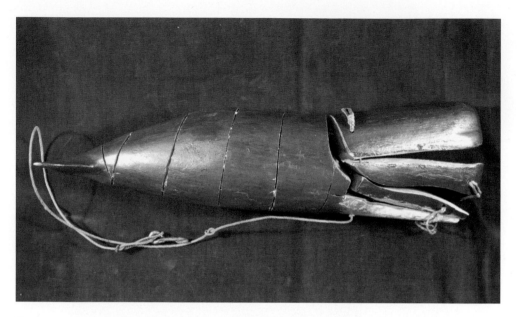

PLATE 13. Pedagogy on the Deck: an articulated wooden model of the cutting-in pattern, probably used for teaching; a nineteenth-century American artifact. Courtesy of the New Bedford Whaling Museum.

PLATE 14. Between the Lines: the log of the ship *Columbia* (early 1840s). Courtesy of the New Bedford Whaling Museum.

PLATE 15. A Sharp Eye on the Watch: marginal illustration from the *Columbia* log, depicting a glimpse of a notched dorsal fin. Courtesy of the New Bedford Whaling Museum.

PLATE 16. Sperm Whale at the Surface: the distinctive bushy, forward blow of *Physeter macrocephalus* (Linnaeus, 1758) has been faithfully depicted (see Plate 14 for provenance information). Courtesy of the New Bedford Whaling Museum.

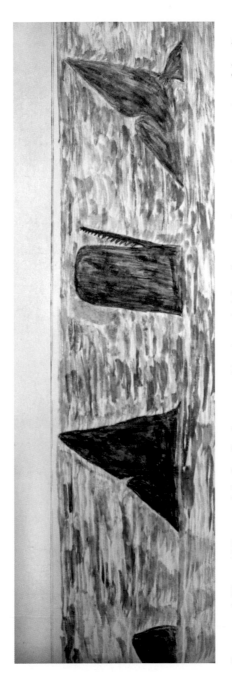

PLATE 17. A Natural History of the Sea Surface: everything in a whaleboat hung on the correct interpretation of the glinting forms that broke, if only for an instant, the surrounding water (see Plate 14 for provenance information). Courtesy of the New Bedford Whaling Museum.

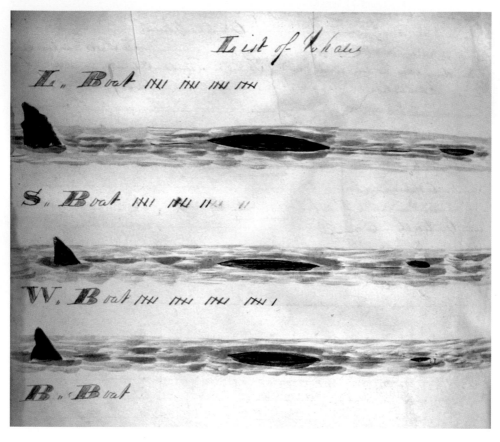

PLATE 18. What Death Looked Like from the Deck: the *Columbia's* tally of kills, by boat (early 1840s). Courtesy of the New Bedford Whaling Museum.

However, while whalemen made their living reducing whales to money via oil, their interest in the animals was by no means merely reductive. When the *Rhine*, out of New Bedford, took a humpback cow in the Caribbean on the 28th of January, 1848, the captain permitted the crew to keep the orphaned calf alongside, alive, drawing it into the harbor of the island of St. Ustacia, for an informal public showing to the inhabitants, before letting it go.[42] And sailors themselves regularly scrutinized the animals with considerable care and often a great deal of imaginative and original thought. It is perhaps not surprising to read that a learned whaler-naturalist like Scoresby—on the strength of his firsthand experience—assumed the right to depart from the cetological classifications of Lacépède, but it is striking to read the zoological speculations of a common-school-educated able-bodied seaman from Avon, New York, named Dean C. Wright, who hypothesized in his journal in the early 1840s that the "class" of humpback whales might conceivably be some sort of hybrid of the right whale and the sperm, since humpbacks shared features of each.[43] Another ordinary whaleman, Joseph Bogart Hersey, took the opportunity of the capture of an anomalous sperm whale on the 21st of February, 1846 (it had visible teeth in the upper jaw, which is uncommon), to record that the species "generally [have] 42 teeth on the lower jaw, whereas this one had 52."[44] He returned to the mouth of his quarry later in the voyage when a strikingly large and old specimen was taken, the majority of whose teeth were rotten. He detailed the behavior of this particular animal, and used the occasion to generalize about this kind of whale:

> Note—the whale when first seen was uncommon long between her spoutings, and apparently feeding, for sometimes she would be down an hour, and at others not more than three minutes. The time which generally elapses from a large sperm whale going down (when not startled) and again rising to the surface is from 45 to 70 minutes.[45]

And Hersey found this very large whale ("120 bbls.," he noted, again using the whaleman's standard metric for measuring both whales and

[42]Kendall Log #394, 29 January 1848, collection of the New Bedford Whaling Museum.

[43]Scoresby dilates on the failings of different taxonomists in Scoresby, *An Account of the Arctic Regions*, vol. 1, pp. 446–450. For Wright, see Stuart M. Frank, *Meditations from Steerage: Two Whaling Journal Fragments*, Kendall Whaling Museum Monograph Series, no. 7 (Sharon, MA: Kendall Whaling Museum, 1991), p. 7. Another clue to whalers' taxonomic ideas about the humpback is afforded by an allusion (Kendall Log # 401, 6 August 1868) to a humpback as a "donkey" in contrast to a "horse" of a sperm whale. Collection of the New Bedford Whaling Museum.

[44]Kendall Log #366, 24 February 1846, collection of the New Bedford Whaling Museum.

[45]Ibid., p. 126.

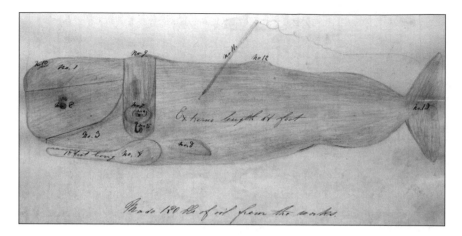

FIGURE 4. A Whaleman Anatomizes His Catch: Joseph Bogart Hersey's sketch of a massive sperm whale, taken by the brig *Phoenix* in 1846. Courtesy of the New Bedford Whaling Museum.

whaleships—barrels) sufficiently remarkable that he took the time to draw it into the log, in a schematic diagram with a key, identifying its parts [Figure 4]. This image merits closer attention, since—like other visual material in the logbooks—it is a helpful source for the whaler's view of the whale, particularly when examined in the context of Hersey's "philosophical" margin comment on the illustrations in his logbook:

> Note: in my drawings of the different fish in this book I have endeavored to give a correct resemblance of those mentioned in my remarks, and would here remark that there is as much diversity in the form and habits of the inhabitants of the ocean as there is in the *shape* and customs of the human family.[46]

This aside takes on special value when it is read after perusing a number of contemporary logbooks, since many of them are illustrated, but by far the majority of their images are tally-stamp depictions of whales used to mark the book's margin when whales were spotted or taken, or to total the take of the ship's different boat-crews (who were engaged in an informal competition throughout the voyage). While some log-keepers drew their own tally-figures in pen, pencil, or even gouache [Plate 8], most used inked wooden stamps [Figure 5], in general at least three: one representing a sperm, another a bowhead (or right, which well into the

[46]Ibid., 14 April 1843; emphasis in original.

1850s were often treated as different varieties of the same creature), and a third representing simply flukes, which indicated whales seen, but not captured. The monotony of the cookie-cutter representations is striking, even when the images are labeled (as they nearly always are) by the number of barrels of oil each captured whale afforded (many of the stamps featured a blank vignetted area where this number could be written in).

Read in the context of this representational uniformity, Hersey's comment on the specificity and individuality of sea creatures is a helpful reminder that, for all the ways that whales were treated as fundamentally fungible entities in the world of whaling (whales, barrels, and dollars were rapidly interconvertible units), there remained an appreciation of the diversity and distinctiveness of the fauna of the sea. The Reverend Henry T. Cheever, who took passage back from mission work in the Pacific via whaleship in the 1840s, and wrote a book about the experience entitled *The Whale and His Captors* (1849), reported that "practiced whalemen" enumerated for him "twelve or fourteen different species of this great sea monster," and indeed a number of logbooks contain "key" pages where numerous cetaceans and other sea creatures are depicted and named, images that almost certainly reflect familiarity with similar images in printed works of natural history [Plates 9, 10; Figure 6] (though the whalemen added their own touches, for instance depicting all the whales spouting blood, as in Plate 10). While relatively few manuscripts reflect so many distinctions as Cheever reported, there are moments when the close knowledge of the whaler speaks through the relatively rigid conventions of the logbook genre. For instance, when Captain James Townsend of the ship *Louisa* came across a dead fin whale floating belly-up in the Atlantic on the 12th of May, 1829, he used his stamp for a right whale in the margin (apparently having no fin whale stamp, since these strong, fast swimmers were not hunted, and were thus more or less irrelevant to the log), but he then meticulously penned in a minute fin on the mid-back of the image, to bring his stamped depiction into conformity with nature and nomenclature.

Returning to Hersey's diagram of the whale [Figure 4] and the key that he wrote to accompany it, several features deserve comment. Most significantly, this image must be recognized as a manuscript instance of a kind of diagram that became an increasingly common feature of published texts on whaling over the course of the nineteenth century: the cutting-in pattern. Because the sperm-whale fishery was a pelagic enterprise—an industry that not only took whales at sea, but also processed them there, melting down the blubber in the brick "tryworks" positioned

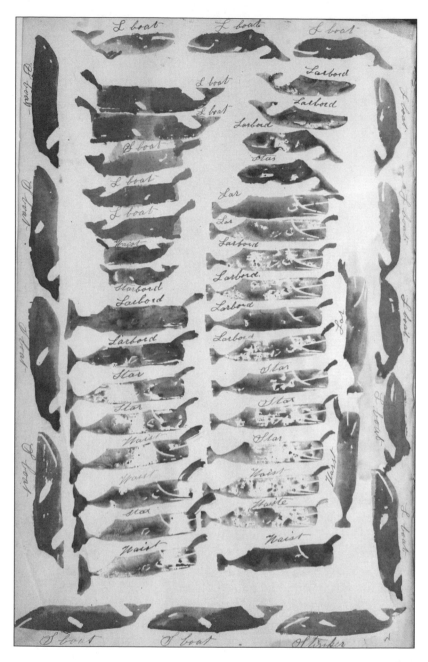

FIGURE 5. Tally Stamps of the *Sarah:* at least six different stamps are used here, and one of the sperm stamps actually shows the cutting lines for the separation of the different parts of the head; the stamp in the upper left corner has been annotated in pen to indicate a pregnant animal (circa 1870). Courtesy of the New Bedford Whaling Museum.

FIGURE 6. Species of Whales: as depicted by the whaleman John F. Martin, aboard the *Lucy Ann* (1843). Courtesy of the New Bedford Whaling Museum.

amidships—a central aspect of the craft was the high-seas butchery of the very large carcasses. This process, described in any work on nineteenth-century whaling, involved lashing the dead whale alongside the ship and then, using tackle fixed in the rigging, winching off strips of the whale's external "blanket" or blubber layer. Whalemen, either placed on the whale itself (and walking on it, like a lumberjack on a log as it rotated in the water), or leaning out over it from platforms suspended over the ship's side, used sharp spades to cut the edges of the strip as the winch peeled it from the body. When the strip was as long as the winch could manage (i.e., the leading edge had been drawn up to block itself—perhaps twenty feet above the water) the men with the spades fixed a new hook at the base of the strip and cut the completed length free, allowing it to be swung on deck for further trimming and mincing en route to the kettles. A fresh strip could then be begun, continuing along from where the old had been cut loose, and spiraling around the animal continuously, exactly like an artfully peeled orange. This process, with a number of attendant operations for processing the head (which contained a large amount of the most valuable spermaceti oil in an interior reservoir, and therefore had to be handled differently), and the jaw (usually severed and lifted aboard so the teeth could be removed), was known as the "cutting-in." It differed in several respects for the major commercial species, since, for instance, bowheads and right whales did not have a reservoir of sperma-ceti in their heads, though they did have the valuable "whalebone" or baleen plates in their mouths, which necessitated different operations [Plates 11 (for sperm) and 12 (for bowhead)].

A close look at Hersey's image with the details of this process in mind reveals that his sketch is more than a drawing of a particular whale. It is in fact a cutting-in diagram, a schematic for this sort of large-scale dismemberment. Below the eye, for instance, at the point labeled with the number 5, the artist has drawn a hook set in a hole. The key reads: "No. 5. Is what is called the blanket piece with the blubber hook in the boarding hole—this is where they generally first hook onto the whale." In other words, the first strip of the peeling process (the strips were known as "blanket pieces") began at the hook as shown, and proceeded up as the yanking tackle turned the animal in the water, the sides of the strip following the parallel lines drawn on the body, circumscribing the eye and running up to the top of the head. Other lines on the diagram indicate the incisions to be made in the careful process of separating the "case" of the head (along the straight diagonal line between regions 1 and 2—Hersey's numeral "2" is that squiggle beside the round ink drip) from the "junk" below. Here, while the basic cuts were standard, different approaches could be used depending on the size of the whale, the conditions of the water, and the habits of the captain and crew. One possibility was to sever the head entirely from the body, and then bring the whole head aboard (if it was small enough—as shown in Plate 11) so as to ensure that none of the best "head matter" would be lost to the waves; another was to use block and tackle to suspend the severed head— bucket-like—over the side, nose-down in the water. From this position the case could be "bailed" by a man standing atop the head and driving a bucket down into the soft and oil-rich tissue.

Whalemen took the niceties of these operations very seriously, so much so that they approached the different options as exacting empiricists: Master John Swift, of the *Samuel and Thomas* out of Provincetown, arranged for a trial of two techniques when his vessel took two similarly sized sperms on 25 June 1847. The log recorded the effort: "cut in the first one by way of experiment leaving the scalp on the body—the other taking the head in whole. The first method proved the most expeditious, though not quite so saving."[47] This meant that each had its virtue: if whales were plentiful in the area, speed mattered more than economy, but rough waters could make the hasty method still more wasteful, and therefore costly.

The earliest known printed cutting-in diagram—much copied during the nineteenth century—appeared in 1798, in Captain James Colnett's *A Voyage to the South Atlantic and round Cape Horn*. Colnett's voyage, in the English ship *Rattler* (he was a Royal Navy officer), was undertaken for the purpose of "extending the spermaceti whale fisheries" into the Pacific, and it received some Crown support. Although Colnett himself was neither a whaler nor a naturalist, he applied himself diligently to his task of reconnaissance, and went so far as to see to it that a small (15-foot) cub sperm whale taken in late August, 1793, off the western coast of Mexico was hoisted on deck in its entirety. He drew the animal, and inscribed on his figure the cutting lines used by the whalemen to butcher it (the *Rattler* had been outfitted by the English whaling firm of Enderby and Sons, and thus had knowledgeable whalemen aboard) [Figure 7]. Colnett also included a detailed key describing the process [Figure 8].

In his useful article "The Historical Evolution of the Cutting-In Pattern, 1798–1867," Michael Dyer has traced the sequence of borrowings and redrawings of Colnett's image, in an effort to establish a

---

[47] Kendall Log #364, 26 June 1847, collection of the New Bedford Whaling Museum.

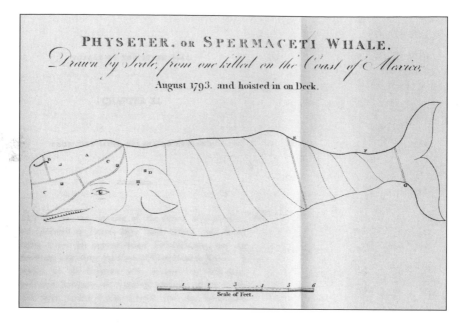

PHYSETER, or SPERMACETI WHALE.

Drawn by Scale, from one killed on the Coast of Mexico,
August 1793. and hoisted in on Deck.

Scale of Feet.

FIGURE 7. The Cutting-In Diagram: Captain Colnett shows how to peel a whale (published in 1798). Courtesy of Firestone Library, Princeton University.

stemma for this iconic figure. Close analysis of a dozen printed exemplars has enabled Dyer to distinguish those changes in the diagrams that can be interpreted as evidence for actual historical developments in the cutting-in process from those alterations most probably owed to sloppy print-piracy [Figures 9–13; Plate 13].[48] Without rehearsing his argument, it suffices here to note that these representations clearly indicate an extensive and minutely detailed anatomical expertise that was the exclusive provenance of the whalemen. It bore little (indeed, perhaps no) relation to the anatomy of the anatomists, but rather functioned as an autonomous domain of natural knowledge. It might even be suggested that something like a "physiology" attended this vernacular anatomy, if of a very particular sort: again observing Hersey's drawing [Figure 4], and focusing on the shaft fixed in the animal's side (labeled "No. 11" in the image), we find that Hersey has annotated this detail, and identified it less as a depiction of the whaleman's fatal tool than as a *pointer*, a blubber-

[48] Michael P. Dyer, "The Historical Evolution of the Cutting-In Pattern, 1798–1867," *The American Neptune* 59, no. 2 (1999): 134–149.

A. *Part of the Head containing liquid Oil, which is covered with a black membrane.* B. *The Spout-hole which runs horizontally along the left side, and is also seperated by the same kind of membrane. The part between the two double lines, is cover'd with Fat of considerable thickness, like that of a hog, these parts make one third of the quantity of Oil the Fish produces, of which the liquid is about one third.* AB. *Part of the Head which of large Whales being too bulky and ponderous to be hoisted on board, is suspended in tackles and the front part cut off as described thus, and the Oil bailed out with buckets; but in small Whales, the head is divided at the double line below* CC. *and hoisted upon deck.* ⬛⬛ *Where the tackles are toggled or hook'd.* D *Where the tackles are first hooked, which is called raising a peice, being thus steadied in the tackles the head is divided at the lowest double line and wore a stern till the fish is flinched, which is done by seperating the Fat from the Body with long handled Iron Spades, as the Whale is hove round by the tackles the Fat peels off, and if any Sea is on the rising of the Ship considerably expedites the business.* E. *A large lump of Fat.* F. *A smaller when the Fish is flinched, or peeled to* E. *it will no longer cant in the tackles, is therefore cut through at the first double line and also at* G. *the Tail being of no value.* H. *The Ear, which is remarkably small in proportion to the body, as is also the Eye from which a hollow or concave line runs to the forepart of the head the Eyes being prominent enables them to pursue their Prey in a direct line, and by inclining the head a little either to the right or left to see their enemy a stern, they have only one row of Teeth, which are in the lower Jaw with sockets in the upper one to receive them, the number depends on the age of the Fish, the lower Jaw is a solid Bone that narrows nearly to a point and closes under the upper, when they spout, they throw the water forwards and not upwards like other Whales except when they are enraged, they also spout more regular and stay longer under water the larger the Fish the more frequently they spout and continue longer under water. The Tail is horizontal with which he does much mischief in defending himself. Their Food, from all the observations I have had an oppertunity of making, has been the Sepia or middle Cuttle Fish. This species of the Whale, is remarkable for its attachment and for assisting each other when struck with a harpoon: and more mischief is done by the loose Fish, than those the boats are fast to, and they frequently bite the lines in two which the struck Fish is fast with. The Ambergrease is generally discover'd by probing the intestines with a long Pole, when the Fish is cut in two at* E.

FIGURE 8. Follow the Instructions: Colnett's description of the cutting-in process. Courtesy of Firestone Library, Princeton University.

hunter's manicule that fingers a relevant anatomical/physiological aspect of the sperm whale.

> No. 11. A lance the point of which has entered the place commonly called the life of the whale, or where a wound would prove mortal.

In this sense the "life" of a whale was a point on its body: a point—*the* point—of greatest relevance to the whaler.[49] Touching the "life" of the whale was the whaler's culminating aim. As if in oblique acknowledgment of this fact, Captain Colnett noted that immediately after he had drawn his specimen, "its heart was cooked in a sea-pye, and afforded an excellent meal."

[49] Davis described the "life" this way: "The blood reservoir, lying under the spine and in the vicinity of the lungs, constitutes what whalemen term the 'life' of the whale. This is described as packed with arteries of great size, coiled in the greatest complexity and containing in their folds an unknown volume of the vital fluid. This is the spot sought for with the keen lance." William M. Davis, *Nimrod of the Sea*, p. 176. Similarly, the New Bedford Captain Francis Post, active in the 1830s and early 1840s, actually hoisted a cub sperm whale on board his vessel so that he and his men might make an anatomical investigation "to observe the position of the seat of life" in order that the harpooneer might "point his lance with a more deadly aim." See Post, "History of the Spermaceti Whale," and other letters from whalemen in Maury, *Explanation and Sailing Directions*, pp. 252–287.

By reference to the prefixed engravings, the following description will be much more readily understood :—

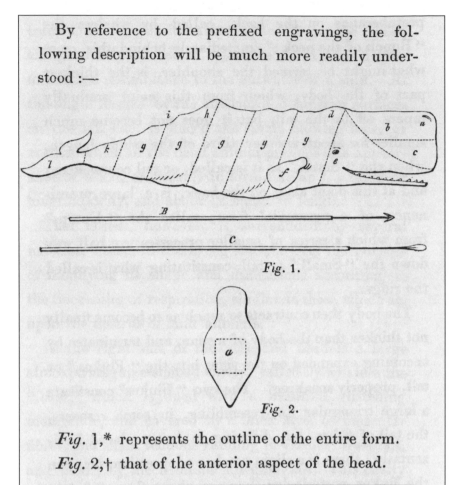

Fig. 1,* represents the outline of the entire form.

Fig. 2,† that of the anterior aspect of the head.

Fig. 1.* *a*, the nostril or spout hole—*b*, the situation of the case, *c*, the junk—*d*, the bunch of the neck—*e*, the eye—*f*, the fin—*g*, the spiral strips or blanket pieces—*h*, the hump—*i*, the ridge—*k*, the small—*l*, the tail or flukes—*B*, a harpoon—*C*, a lance.

Fig. 2.† *a*, the lines forming the square are intended to represent the flat anterior part of the head.

FIGURE 9. Thomas Beale's Cutting-In Diagram: along with a small head-on schematic view of a sperm whale, showing what Melville would later call "The Battering-Ram" (image published in 1835). Courtesy of the New Bedford Whaling Museum.

To show how this process is conducted, and as exhibiting a faithful outline, taken by measurement, of a young individual of this genus, we refer to the accompanying cut copied from a sketch of Colnett's.

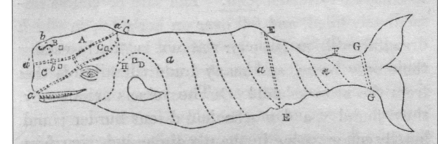

The tackles are hooked at D ; *a, a, a,* are spiral stripes successively removed ; when removed as far as E, the carcase will no longer "cant" in the tackles, and it is therefore cut through at the line E, E, and also at G, G, the tail being of no value. The compartment A shews the part of the head which contains the liquid oil. Being suspended by the tackles, the front part is cut off at *b, b,* and the oil baled out with buckets. When the whale is small, the head is divided at the line *c', c',* previous to its being hoisted on deck ; the space between *a', a',* and *c', c',* also contains much oil. B is the

FIGURE 10. Robert Hamilton Copied from Colnett: but made the whalemen's anatomy available to a large readership of landlubbers, via publication in Jardine's *Naturalist's Library* series (1837). Courtesy of Firestone Library, Princeton University.

But if cutting-in images are valuable sources in the whalemen's natural history—in that they provide clear documentation of what whalers knew about the anatomy and physiology of their prey, and indeed evoke a broad domain of tacit knowledge about the body and vitality of these animals—it must nevertheless be emphasized that this was, in a literal sense, a "superficial" anatomy. It is surprising, for instance, to read Captain Scoresby (whose hands had rifled the carcasses of more than one thousand bowheads in the Arctic seas) cite a bookish authority for the number of ribs the animal possessed.[50] How was it that Scoresby did not

[50] Scoresby, *An Account of the Arctic Regions,* vol. 1, p. 463.

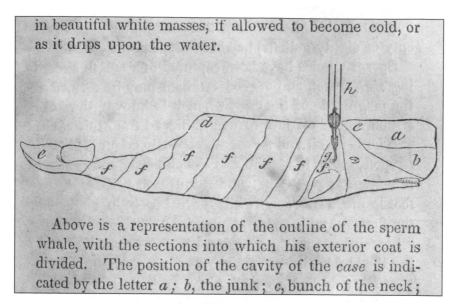

in beautiful white masses, if allowed to become cold, or as it drips upon the water.

Above is a representation of the outline of the sperm whale, with the sections into which his exterior coat is divided. The position of the cavity of the *case* is indicated by the letter *a ; b*, the junk ; *c*, bunch of the neck ;

FIGURE 11. Francis Allyn Olmsted's Cutting-In Diagram: showing the block and tackle positioned to winch the first strip (1841). Courtesy of Firestone Library, Princeton University.

know this himself, from his own extensive experience? The answer, of course, is that even a committed naturalist-whaleman like Scoresby had never, when engaged in business, taken the opportunity of the hunt to perform an exhaustive dissection on these animals as they floated beside his ship. In fact, given that the fat-stripped carcass not infrequently went directly to the bottom (when no longer buoyed by its insulating blubber layer), such an undertaking might have been impossible—putting aside how unpleasant it would have been to try, standing on a cadaverous raft waist-deep in a slurry of polar water and viscera. While Scoresby had once or twice seen to it that a stomach was opened (to observe what the animals had been eating), he was forced, for all his intimacy with the creature, to rely on the published work of museum naturalists for the deep internal anatomy of the *Balaena mysticetus*, particularly the skeleton. Captain Fish, on the stand in *Maurice v. Judd*, found himself obliged to make the same confession: while he knew sperm whales up close, and had cut into many dozens of them, and had, with his crews, pawed through their intestines (as whalemen regularly did in pursuit of lumps of precious ambergris), he had to admit that the bones lying under their meat were beyond his ken; when Sampson pressed him on the internal

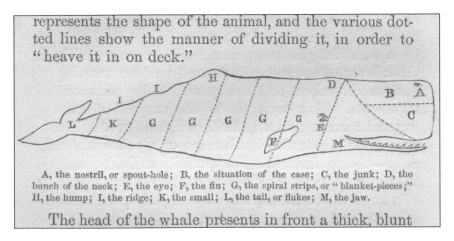

represents the shape of the animal, and the various dotted lines show the manner of dividing it, in order to "heave it in on deck."

A, the nostril, or spout-hole; B, the situation of the case; C, the junk; D, the bunch of the neck; E, the eye; F, the fin; G, the spiral strips, or "blanket-pieces;" H, the hump; I, the ridge; K, the small; L, the tail, or flukes; M, the jaw.

The head of the whale presents in front a thick, blunt

FIGURE 12. Chase and Heaton Copied from Beale: and gave their whale a particularly fierce-looking "eyebrow"; one suspects that whalemen would have snickered (1861). Courtesy of Firestone Library, Princeton University.

anatomy of the whale's fin, Fish acknowledged that "I have never dissected any of them," and Sampson shot back, "I can very well perceive it Mr. Fish."[51]

So cutting-in, despite all the gargantuan surgery it entailed, was not "dissection."[52] At the same time, the whaleman's "superficial anatomy" must be acknowledged (perhaps paradoxically) as a profound knowledge of the superficies of the animal. The "blanket" was what mattered to whalers, and their cetological nomenclature reflected their preoccupation with the three-to-fifteen-inch-thick blubber envelope that was their livelihood. For instance, whalers recognized the "dry-skin whale" as a type of cete, a designation that passed unnoticed among taxonomists: a "dry-skin" whale was a whale of any species that, for one reason or an-

[51] *IWF,* p. 19.

[52] Though the term was used in printed sources: "The 'cutting-in' of a large whale is truly a formidable undertaking. It is surgery, or dissection, on a gigantic scale" (William M. Davis, *Nimrod of the Sea,* p. 77); "The internal anatomy of the whale is to me a subject of great curiosity, and I wish it were in my power to report a full and accurate, leisurely *post-mortem* of the subjects we have discussed. But a few clinical notes, roughly taken by the bed-side, as the whalemen have been operating between wind and water with their professional spades and lances of dissection, are all I have to exhibit" (Cheever, *The Whale and His Captors,* p. 78). In the early 1840s, as the industry reached its peak strength, there were several formal dissections of large whales conducted in the United States. See, for instance, J.B.S. Jackson, "Dissection of a Spermaceti Whale and Three Other Cetaceans," *Boston Journal of Natural History* 5, no. 2 (1845), pp. 137–171.

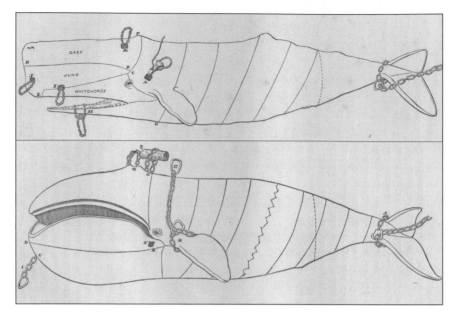

FIGURE 13. A Superficial Anatomy: Charles M. Scammon, an American West Coast whaleman-naturalist, offered the most detailed cutting-in diagrams of the nineteenth century, depicting both a sperm and a whalebone whale (1874). Courtesy of Firestone Library, Princeton University.

other, had little or no fat upon its body. Such a characteristic was, among whalers, salient enough to eclipse the morphological differentia of the schools. Similarly, whalemen had several idiosyncratic names for specific regions of a whale's external anatomy, and yet this terminology did not denominate any visible "part" of the animal that might be spotted by a passive observer, but rather indicated the way whalers perceived the *quality* of the skin at those points under particular conditions: Dean C. Wright pointed out that in sperm whales "Both the case and junk are guarded [by] a substance called headskin, which is very hard, and is almost impenetrable to a harpoon, and thus their head is rendered very formidable in their defence"; those who pursued right and bowhead whales were accustomed to talk of "slack blubber," an impenetrable area that the whales could move about their bodies at will, by twisting themselves to thicken their external tissue.[53] Tellingly, the term "black-skin" could be used among whalers as a synecdoche for the whale as a whole.

In these details, in the extensive vocabulary and metrics applied to

[53] See the comments by Dean C. Wright in Frank, *Meditations from Steerage,* p. 5. Hohman discusses slack blubber: Hohman, *The American Whaleman,* p. 180.

the blanket, and in the evolving craft of the cutting-in process, we see the contours of the whaleman's "superficial" anatomy, an anatomy that was superficial only in the narrowly literal sense: there was nothing "shallow" about it (excepting the depth into the body it reached), since, in its domain, it was in every respect more detailed and closely observed than anything zoologist-savants had to offer. At the same time, it is worth noting that the contrast of this whaler-anatomy with the preoccupations of the new comparative anatomy—given the latter's increasing attention to deep inner structure—could not have been more complete.[54]

The notion that the whalemen possessed a superficial natural history felicitously evokes another significant dimension of whalers' knowledge of nature. Spending time with whaling logs, one is promptly struck by the many sea-surface views they offer—depictions of the water, of the "face of the deep," abound, sketched between lines of text or brushed into narrow margins. What takes a while to appreciate is the way that these images constitute a veritable natural history of the surface of the ocean. Unlike cabinet students of nature, who received their sea specimens in jars or crates, or perhaps encountered their large cetaceans distended on the strand, the whaleman's visual contact with sea fauna—the different species of shark, turtle, dolphin, whale, and fish that were his continuous preoccupation for hundreds and often thousands of days— was most often a glimpse at the juncture of water and air. These men saw what *surfaced:* a fin, a tail, a nose, a spout. It is little wonder then that their journals represent the "fish out of water"—the image of a sea creature lying on the page—much less often than they depict the sightings relevant to their working world: the fin, blow, or splash that breached the undulating plane separating sailors from the deep. The keeper of the log of the *Columbia,* for instance, painted dozens of sea creatures, consistently limiting himself to the bit of black that tipped out above the waves [Plates 14–17]; his sketches amount to a field guide for the naturalist in the crow's nest, an identification key to the whales one saw as one sailed and sought them where they lived. Indeed, so taken was this log-keeper with this superficial view of the sea and its inhabitants, that even when he tallied the takings of the ship's four boats, he depicted rep-

[54]Melville in fact seized on the great difference between the external and internal anatomy of the whales in chapter 55 of *Moby-Dick,* where Ishmael cites Hunter: "In fact, as the great Hunter says, the mere skeleton of the whale bears the same relation to the fully invested and padded animal as the insect does to the chrysalis that so roundingly envelopes it. This peculiarity is strikingly evinced in the head, as in some part of this book will be incidentally shown. It is also very curiously displayed in the side fin, the bones of which almost exactly answer to the bones of the human hand, minus only the thumb." See also chapter 86: "Dissect him how I may, then, I go but skin deep; I know him not, and never will."

resentative dead whales, lying in water, breaking the surface only in three places, three distinctive darts of black-skin that were the "footprint" of the whale in the world of daylight [Plate 18]. Since "reading" the surface of the sea was the primary means by which whalemen found, identified, and tracked whales, it is perhaps no surprise to discover such care lavished on this form of silhouetted and glancing natural history. As Browne put it, concisely, "it is the primary object in whaling to see whales when they appear above the surface of the water, so it is the chief qualification of a good whaleman to understand thoroughly the different species of whales, and how to distinguish them."[55] In an effort to show what this meant, he included a practicum of whaler-knowledge: a depiction of the blinkered view of the three most familiar species [Figure 14]. Blinkered, to be sure, but by no means blind; if anything, this was a vision enormously refined and focused in its calculus of the flashing and the fragmentary. There was little patience for whalemen who failed this test: lowering the boats for finbacks, say, mistaking them for a catchable species, wasted everyone's time and energy, yet it was certainly known to happen. The anonymous log-keeper of the *Cowper*, out of New Bedford for sperm whales in the summer of 1846, showed little mercy for a crewman inept in this art: "saw whales lowered and the second mate could not tell which end the whales [*sic*] head was on, so ends."[56] A mid-nineteenth-century French naturalist who spent some time among whalemen was flabbergasted by their abilities when it came to reading the splash and spray:

> The whaleman knows how to distinguish the true whales [right whales] from every other species of cetacean by the form and by the color of whatever appears above the surface of the sea when they are swimming, as well as from their manner of rising and diving; and at a greater distance by the shape of the vapor from their exhalations, which from a certain distance resemble (indeed, could be taken for) jets of water. Even in the dark of night, he can distinguish them from the mere sound of their breath, which can sometimes be discerned more than a kilometer off.[57]

For all the significance of these two kinds of "superficial" natural history (the surface of the whales and the surface of the sea), whalers were

[55]J. Ross Browne, *Etchings of a Whaling Cruise*, p. 163.

[56]Kendall Log #64, p. 66, collection of the New Bedford Whaling Museum. "So ends" was a common phrase for closing an entry; it means "so ends the day."

[57]Daniel Fredrik Eschricht, "Sur une Nouvelle Méthode de l'Etude des Cétacés," *Comptes Rendus des Séances de l'Académie des Sciences* 47 (1858), pp. 51–60, at p. 53.

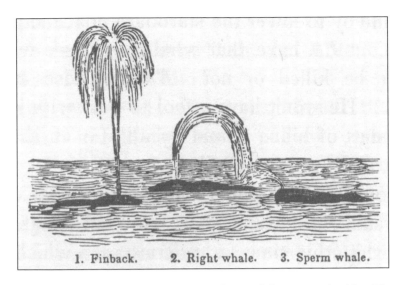

1. Finback.   2. Right whale.   3. Sperm whale.

FIGURE 14. Reading the Mist: J. Ross Browne's key to species identification, according to pattern of blow (1846). Courtesy of Firestone Library, Princeton University.

possessed of a penetrating intimacy with their prey as well. Colnett's hearty "sea-pye" reminds us of an important aspect of the whaleman's "communion" with the animals he hunted: more or less perpetually hungry (and often actually undernourished), on account of their monotonous diet of poor salt meat and hard biscuit, whalers seldom spurned the "sea beef" afforded by the cetes.[58] Porpoises were a perpetual favorite (harpooned from the bowsprits), and the liver and lean muscle were eaten with relish, often in floured dumplings known as "porpoise balls"; a similar dish could be concocted out of the brain of sperm whales, kneaded into "fritters."[59] The thin edge of the lip of the right whale, allowed to simmer overnight in the hot oil of the trying kettles, "becomes the consistence of a jelly, which, when eaten with salt, pepper, and vinegar, closely resembles pigs' feet."[60] Blubber could be pickled, and portions of the tail could be parboiled and fried, and Scoresby acknowledged that

[58]"Sea beef" was a common term. For examples of usage: Cheever, *The Whale and His Captors*, p. 97; William M. Davis, *Nimrod of the Sea*, p. 18.

[59]On eating "porpess," see Kendall Log #364, p. 293, collection of the New Bedford Whaling Museum. See also: J. Ross Browne, *Etchings of a Whaling Cruise*, p. 63; and Hamilton, *Natural History of the Ordinary Cetacea*, p. 226. For a general study of the question, see Nancy Shoemaker, "Whale Meat in American History," *Environmental History* 10, no. 2 (2005), pp. 269–294.

[60]William M. Davis, *Nimrod of the Sea*, p. 390.

"the flesh of young whales, I know from experience, is by no means in-different food."[61] It is a telling clue to the whalemen's sense of how the "spouting fish" fell taxonomically with respect to the quadrupeds that the term "cow-fish" was used to designate a host of the smaller odontocetes, and while one might readily talk of "whale bacon," fish bacon was inconceivable.[62] Against this background Melville's highly wrought depiction of "Stubb's Supper" (chapter 64 of *Moby-Dick*), with its evocation of the man-shark devouring his prey as a steak while the shark-sharks work cannibalistically on the carcass below him, must be understood as an allegorical rendering of a familiar shipboard ritual.

Such rites of blooding and consumption represent more than just appetite. They also elaborate the charged intimacy of the huntsman: it was the task of skilled whalers to develop an inwardness with the ways of the whale, to grow familiar with the sensory range, sight lines, foibles, and wiles of their prey, and to achieve reliable intuitions about the beings they pursued. "It smells rather whaley," noted the log-keeper of the *Columbia*, cruising the line on the 11th of January, 1844, attesting to the hunter's keen sixth sense as he took in the water conditions, the birds and flotsam and glinting surface activity of the equatorial Pacific; he added, as a note to himself and the crew, "look sharp boys."[63] Looking sharp for whales meant honing one's own horizon-eye vigilance, for certain, but it also meant paying very close attention to the perspective of the prey itself. Illustrating his account of the chase, William Davis included an image of a sperm whale's visual field, and showed how a boat might approach from the rear without "gallying" (spooking) the catch [Figure 15]. Whaling lore included intricate ideas about the whale's sensory universe, some of which may have functioned as mnemonic devices or rules of thumb for the assault. For instance, many American whalers held that sperm whales were sensitive to the passage of a pursuing boat through their "glip," or the smooth streak of water in their immediate swimming

[61] Scoresby, *An Account of the Arctic Regions*, vol. 1, p. 476. The powerful whaling agent Charles W. Morgan (who never actually went on a whaling voyage but was a central figure in the New England industry) even decided, when the opportunity presented itself, to taste the creature on which his fortunes were made. He noted that a number of Nantucket tables were furnished with whalemeat when it was on hand and fresh. See "Address before the New Bedford Lyceum," Charles Waln Morgan Papers, 1817–1866, box 7, series Y, p. 6, collection of the New Bedford Whaling Museum. While he was visiting the island (in the 1820s) a whale was caught near shore and towed into the harbor: "Curiosity not only tempted me to visit the vessel, but to take a slice of the delicate contents of the try pot, which with a piece of buiscuit [*sic*] (*and a good appetite*) I did not find altogether unpalatable"; emphasis in original.
[62] Wilson, *Cruise of the "Gipsy,"* p. 155–156.
[63] Kendall Log #213, p. 66, collection of the New Bedford Whaling Museum.

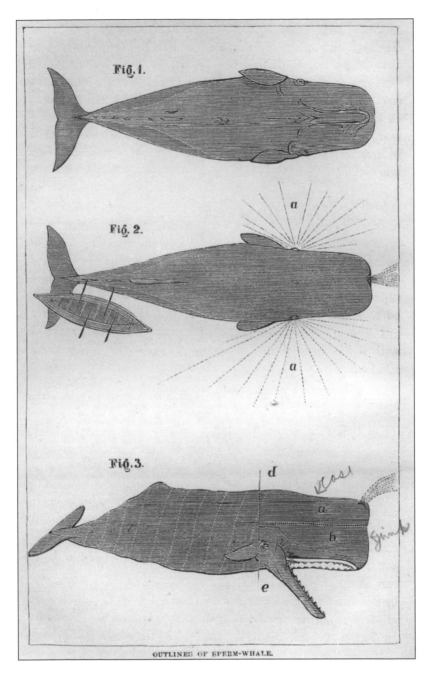

FIGURE 15. The Line of Attack: William M. Davis considers the sperm whale's visual field (1874). Courtesy of Firestone Library, Princeton University.

wake. It is not clear how this strange idea arose, redolent as it is of sympathetic magic, but application of the principle would have resulted in a shallow angle of approach from the rear—an excellent way to avoid both the animal's visual field, and its flukes, which could be used for a fearsome defense.[64]

The different means by which the various species reacted to attack, and their respective vulnerabilities, these were the most essential elements of whalemen's whale-knowledge, and they carried over into the kind of anthropomorphic preoccupations not uncommon among those who make their livelihoods outsmarting wild animals. Dean C. Wright, for instance, who served for four years as a harpooneer, and thus went eye to eye with hundreds of whales in three oceans, confided to his journal that sperm whales were distinctive not only in profile, but also in their intelligence:

> a sperm whale is a species alone, no other kind seeming to be of his form or nature, for he is not only a different shape from all other whales and worth more than any others, but the sailors say that "they know a d—n sight more than the others"; and I think there is some truth in the expression.[65]

Moreover it was recognized that some individual whales were "smarter" than others, a theme difficult to avoid after 1821, when a suite of articles and pamphlets recounted to chilled readers the tragic tale of the wreck of the whaleship *Essex,* sunk by a bull sperm whale on the equator in the Pacific in November of 1820. Whales crunching twenty-five-foot open whaling dinghies was a standard hazard of the chase; a whale using its head to ram to destruction a fully rigged *ship*—hulking with its multiple decks and masts and armed with a hull designed to stave off Cape Horn tempests—was a sensation, and countless readers devoured the survivor's tales (which included lurid details of murder and cannibalism), and shivered to read the eyewitness accounts of the injured animal's first blow to the bow, and then its retreat, pause, and second fatal charge—fatal, that is, to the *Essex.* Such a story, so widely hailed, gave immediate resonance to whalemen's stories about specific whales that were "wild," or, as the terminology had it, "ugly."[66]

---

[64]Hohman and Davis both discuss the "glip": Hohman, *The American Whaleman,* p. 158; William M. Davis, *Nimrod of the Sea,* p. 181.

[65]Dean C. Wright, cited in Frank, *Meditations from Steerage,* p. 6.

[66]Some of these whales earned names, most famously, Mocha Dick, the "real" Moby Dick, who gained recognition when J. N. Reynolds composed a brief news report on his marauding in the 1830s (see Jeremiah N. Reynolds, "Mocha Dick or the White Whale of the Pacific," *Knickerbocker Magazine* 13 [May 1839], pp. 377–392). For a whaleman's view of the mean-spirited cetacean, see the chantey "All the Whales are Wild and Ugly" by George Edgar Mills,

The social behaviors of the animals mattered as much as their individual wiles, particularly as such habits might be exploited in the hunt. Hersey, for instance, mentioned in his log that the boats of the *Phoenix*, coming upon a group of sperm whales that included several females with calves, tried the whaleman's familiar trick of striking a cow whale and just hanging on, in the hopes that the other females in the pod would come to her aid (or "gam" with her, as sailors put it)—creating excellent opportunities for multiple catches.[67] Sometimes sticking a calf worked even better, and the practice—used in the pursuit of sperm, right, and bowhead (and later in the century for gray whales as well)—struck many observers (and even some whalemen) as unfortunately cruel, particularly when animals were thus betrayed by what looked like "maternal affection."[68] Dean Wright thought too much was made of this tangibly un-fishy characteristic of his prey, and he noted that "The general idea of the great affection which the cow has for her calf is rather exaggerated if my experience be correct, for I have frequently seen the cow leave her calf; but in some cases she shows great regard for her young and will rather stay and be killed than leave it."[69]

Stories of both flight and fight could serve to "humanize" these creatures, so alien to most non-whalers. A number of whalemen who commented on whales for land-bound readers went out of their way to emphasize that, mothers aside, whales were generally "timid" animals, easily frightened and quick to take flight, and as the global whaling industry climbed to its apogee at mid-century (and as the depletion, indeed the potential extermination, of great whales became a topic of commercial

---

dated 1855 (Frank, "Ballads and Songs of the American Sailor," #158). There is a very large literature on the *Essex*, including, recently, the popular work by Nathaniel Philbrick, *In the Heart of the Sea: The Tragedy of the Whaleship* Essex (New York: Viking, 2000). The narratives of Owen Chase, Thomas Chappel, and George Pollard remain in print: Owen Chase et al., *Narratives of the Wreck of the Whale-Ship* Essex: *A Narrative Account* (New York: Dover, 1989). I came across a manuscript account of a sperm whale "attacking" a ship, the *Joseph Maxwell* of Fairhaven, on 16 October 1845 (see Kendall Log #124, p. 99, collection of the New Bedford Whaling Museum). The struck whale "hove his junk out struck the ship on the starboard bow," as the crew "put in one lance and to [*sic*] irons." The whale subsequently "came up a stern spouting blood" but apparently escaped to windward.

[67] This is widely noted in the published and manuscript materials on whaling. For a manuscript discussion, see Kendall Log #366, circa p. 140, collection of the New Bedford Whaling Museum.

[68] Whales' willingness to stand by their young attracted commentary very early in the history of the industry, and was immortalized in English by the seventeenth-century poet (and member of the Royal Society) Edmund Waller in "The Battle of the Summer Islands," which depicts the confrontation between the inhabitants of Bermuda and two whales (mother and calf) stranded in the shallows of the island. The poem was a regular reference for whaleman-authors in the nineteenth century.

[69] Dean C. Wright, cited in Frank, *Meditations from Steerage*, p. 6.

and *bien-pensant* concern), several authors hazarded whale's-eye views of
the world, often tinged with maudlin sentimentality: bookshelves in 1849
held a volume that claimed to be nothing less than "The Whale's Biography," and a year later *The Friend* of Honolulu—a newspaper published
by the Oahu Bethel Church, and perhaps the most widely circulated pe-
riodical among Pacific whalemen—printed a plaintive "letter to the edi-
tor" from "Polar Whale" (address: "Anadir Sea, North Pacific") in which
this spokesman for an "Old Greenland family" lamented the fate of his
many confreres, and pleaded for "friends and allies" to "arise and revenge
our wrongs" lest ignominious extinction descend upon his "race."[70]

Such bathetic treatments undermined—or perhaps simply remained
in tension with—the characterizations of whales as "monstrous" beings,
depictions that retained currency in much nineteenth-century popular
published writing on the cetes, and that had a place in the sea chanteys
and doggerel of the whaleship as well. It took a minister like Cheever to
truck out Milton (who notoriously likened Satan, floating on the burning lake of hell, to the vast bulk of the Leviathan), but New England
whalemen throughout the century joined in choruses of "A Whaling
Song," with its invocation of the "monsters of the deep," "the monstrous
fish" with her "monstrous body," who was the whaler's "deathful prey."[71]
And whalemen did not shirk a proper Miltonian diabolism of their own:
Cheever might go so far as to assert that the "moving sea-god" was nothing less than the embodiment of Mammon himself (particularly since
seamen's relentless pursuit of blubber-lucre generally meant that there
were, as the saying had it, "no Sundays off-sounding," i.e., no Sabbath at
sea), yet whalemen too hypothesized that "wild and ugly" whales had the
actual devil in them.[72] Moreover, whalers (and not just Melville) sometimes took up a gallows humor, and embraced the diabolism of their

[70] "The Whale's Biography" is part of the subtitle of Cheever, *The Whale and His Captors.*
The letter appears in *The Friend* (Honolulu), 15 October 1850, pp. 82–83. The point about timidity is widely made in published and unpublished sources. See, for instance, "Address before the
New Bedford Lyceum," Charles Waln Morgan Papers, 1817–1866, box 7, series Y, pp. 3 and 17,
collection of the New Bedford Whaling Museum.

[71] For the Milton image, see Cheever, *The Whale and His Captors,* p. 64. For "A Whaling
Song," see Stuart M. Frank, "Classic American Whaling Songs," unpublished manuscript, p. 2.
The lyrics of this song date to the mid-eighteenth century, but texts are preserved throughout
the nineteenth.

[72] Cheever, *The Whale and His Captors,* pp. 248–249; for whalemen singing of diabolic
whales, see the "Bowhead Whaling Song" (date not established, but not later than 1872 and
probably much earlier), verse 16: "For Bow-Heads are like spirits though once they were like
snails / I really believe the devil has got into bowhead whales." Frank, "Ballads and Songs of the
American Sailor," #157.

trade as a whole.[73] Certainly there were pious whalemen, like Clothier Pierce, who scribbled in their logs ejacular supplications for divine aid ("Forgive me Heavenly Parent, my past Sins and transgressions & Oh, Lord Bless us soon to get One Whale is my Earnest Prayer"), and there were those who moralized the whaleman's role as a godly alchemist who converted "creatures of darkness" into the sources of domestic light, but William Davis took a wry look at such energies: he suggested that a whaleman might return home, forever the most prodigal of sons, and announce simply, "Father, I have whaled," an admission that was the résumé of all sin.[74] This vision of the whaleship as an impious saturnalia is perhaps best captured by the lyrics to "Jack and the Whale," the American whaleman's scurrilous inversion of the tale of Jonah, which was passed along and altered by oral transmission through the nineteenth century: The blubber-hunter Jack, confronting "the whale that swallowed Jonah," finds himself similarly ingested, but in defiance lights up his pipe and puffs away, making the beast of divine retribution sick to its stomach (a very different sort of "harrowing of hell"). When the whale tries to spit him up, Jack refuses to go, and, clinging in the monster's mouth, proceeds to kill the whale with his bare hands, tossing *it* ashore and cashing in comfortably on the value of the oil.[75] So much for divine retribution.

Somewhere in the mixture of devilish whales and apostate whalers, a subtle kinship of outlaws could take shape. The log-keeper aboard a New Bedford ship recorded a battle royale against a large sperm on the 16th of April, 1849, with a tip of his cap:

> saw whales and lowered and struck one . . . he was a real devel, after being fast to him for about four hours he went off by tak-

[73] Melville is, of course, relentless in his depiction of the satanic world of outcast men, bent on death, steeped in blood, stirring boiling kettles of oil stoked by greasy blazes (the tryworks were fueled by the "cracklings" from the pot, the bits of oil-soaked skin and flesh). For lurid instances of his stygian mode, see chapter 95, "The Cassock," and chapter 96, "The Try-Works." A recent vein of literary criticism and maritime history (departing from the work of Marcus Rediker and Michel Foucault) has been much taken with the ship as a "heterotopia," a world of inversion, saturnalia, and subversion, a notion that finds choice textual supports in the whaling industry. For a taste of this literature, see Cesare Casarino, *Modernity at Sea: Melville, Marx, Conrad in Crisis* (Minneapolis: University of Minnesota Press, 2002). Not irrelevant in this respect is the chilling motto of the early twentieth-century Antarctic whalers, cited in J. N. Tønnessen and Arne Odd Johnsen, *The History of Modern Whaling* (Berkeley: University of California Press, 1982): "Beyond forty [degrees South latitude] no law, beyond fifty no God!" (p. 158).

[74] Clothier Pierce cited in Frank, *Meditations from Steerage,* p. 26. William M. Davis, *Nimrod of the Sea,* p. 34.

[75] See Frank, "Classic American Whaling Songs," p. 10.

ing the line . . . there was attached to him 12 irons 4 lances two
lines and a half, 2 drugs [wooden sea anchors used to slow up a
fleeing whale], one spade, and he capsized the starboard boat.
success to him I say.[76]

Such an aside points to the ways that the whaleman's familiarity with his
quarry might limn a sympathetic union stranger and more profound
than the aromatic special pleading of the mission press with its saccha-
rine pleas for the persecuted "Polar Whale." A hunter might sing out—
as whalemen did—"Death to the living, long life to the killers" as a way
to speed the oar in pursuit of "wood on black-skin" (running the boat
"aground" on the fleeing whale's back, to give the harpooneer point-
blank range), but the same phrase also functioned to help the killers keep
a certain kind of distance.[77] After all, both Cheever and J. Ross Browne,
for instance, cited whalers expressing misgivings about the sufferings of
their catch, and while Browne may well have given himself fictional license when he made those misgivings the theme of a passage that re-
counts the whaleman's ultimate nightmare, this does not detract from
the creepy power of "Barzy's dream" (about the "un-Christian business"
of whaling): "I dreamp," announces Browne's protagonist to his messmates, roused from their berths by his cries, "I dreamp I WAS A WHALE."
And Barzy proceeds to narrate his experience of "being" the object of
the ship's relentless pursuit, detailing his bloody capture, and his terrifying awareness of each stage of his rendering—as he is skinned, dis-
membered, and, finally, melted, cognizant of his own unmaking all the
way through the stinking process. Barzy summed up the lesson thus:
"Whales has feelings as well as any body. They don't like to be stuck in
the gizzards, and hauled alongside, and cut in, and tried out in them 'ere
boilers no more than I do."[78] His fellow whalers declare this unsettling
reverie "the climax of all the dreams we had ever heard," but as men
sealed in a predatory pact—and many of them hailing from farming
lives in which slaying large animals was a familiar autumn labor—they
shake off its uneasy implications.[79]

[76] Cited in Mary Malloy, "Whalemen's Perceptions of 'The High and Mighty Business of
Whaling,'" *The Log of Mystic Seaport* 41, no. 2 (1989), pp. 56–67, at p. 61.
[77] The motto is a not-uncommon scrimshaw epigram; see Frank's annotations on the journal of John Jones: Frank, *Meditations from Steerage*, p. 24.
[78] J. Ross Browne, *Etchings of a Whaling Cruise*, pp. 198–200. Cheever cites this passage:
Cheever, *The Whale and His Captors*, p. 146.
[79] To contextualize these sentimental considerations properly, it is necessary to consider
broader changes in attitudes toward pain in general, and animal pain in particular, in the English-speaking world across the long nineteenth century. The points of departure here are

Even if we dismiss Browne's hallucinatory tale as a confection, it cannot be denied that whales and other sea creatures had a way of worming themselves into the minds of the men who thought of little else for years on end. William W. Taylor, second mate aboard the *South Carolina* out of Dartmouth, edging along the east coast of Greenland in January of 1837, recorded an oblique account of a "remarkable dream" about walruses and other beasts of the polar waters. And as the hardscrabble sailor Daniel Kimball Ritchie—recently sprung from jail in the Sandwich Islands—found his slow way home on the *Israel*, meandering across the widest part of the Pacific in pursuit of sperms, he realized the beasts had come to occupy the whole of his mental universe: "Cruising for sperm whales but in vain. . . . I am in day looking for them & in night thinking of them."[80]

It is also the case that the language of whalemen contained a cluster of terms whose slippery usage betrays that something more than a stout manila line bound whalers to their prey, terms that suggest sometimes the resurrection of a dead metaphor can bring a buried kinship to light. This went beyond the curious convergence of slang like the noun-and-verb "gam," which applied both to gregarious whales in their groupings and gregarious whaleships drawn up at sea for sociability. And perhaps not too much should be made of the fact that whales and whaleships shared the common denominator of the cooper (since both were forever measured, as discussed above, in barrels). But a close reading of the words that whalers left reveals elaborate and suggestive puns that lurked in the whalemen's repertoire—for instance, the use of the phrase "her whaleship" (by way of "her worship") to name not a vessel but a cete, as in the pregnant pre-Freudian chantey "How to Catch a Whale":

> On land Poor Jack secured his prize,
> He stuck a knife in her belly,

Keith Thomas, *Man and the Natural World: Changing Attitudes in England, 1500–1800* (Oxford: Oxford University Press, 1983); Turner, *Reckoning with the Beast*; Ritvo, *The Animal Estate*; Anita Guerrini, *Experimenting with Humans and Animals: From Galen to Animal Rights* (Baltimore: Johns Hopkins University Press, 2003), and Nicolaas A. Rupke, ed., *Vivisection in Historical Perspective* (London: Croon Helm, 1987). There are also various specialized studies that are useful, including: Paul White, "The Experimental Animal in Victorian Britain," in *Thinking with Animals*, edited by Daston and Mitman, pp. 59–81; and Andreas-Holger Maehle, "Cruelty and Kindness to the 'Brute Creation': Stability and Change in the Ethics of the Man-Animal Relationship, 1600–1850," in *Animals and Human Society: Changing Perspectives*, edited by Aubrey Manning and James Serpell (London: Routledge, 1994), pp. 81–105.

[80] Kendall Log #411, 7 January 1837; Kendall Log #478, circa 20 May 1845. To be fair, Ritchie spent a good deal of time longing for the American West as well, and visions of homesteads regularly crowded on his visions of sperm whales. Both volumes are held in the collection of the New Bedford Whaling Museum.

> Which made her whaleship roll her eyes,
> And look like a lump of jelly . . . [81]

Inside the body of his catch Jack finds a fleet of actual whaleships, which sail stately out of her corpse: his prize was thus carrying whaleships in her hold just as whaleships carried their whales, stowed down in their casket-kegs. Such phantasmagoric folklore traditions leave little doubt that "whale-men" were, we might say, acutely unconscious that theirs was a compound name. Jeremiah N. Reynolds, coming upon an exemplary American whaleman on a New York vessel in the Pacific in the early 1830s, gave clear voice to these zoomorphic whisperings:

> Indeed, so completely were all his propensities, thoughts, and feelings, identified with his occupation; so intimately did he seem acquainted with the habits and instincts of the objects of his pursuit, and so little conversant with the ordinary affairs of life; that one felt less inclined to class him in the genus *homo*, than as a sort of intermediate something between man and the cetaceous tribe. [82]

For confirmation that there was something, well, "amphibious" in the progressive convergence of the dedicated whaleman and his prey, we need look no further than the durable nautical slang for a whaler-seaman, who might be known to chums as a *fish*. [83]

Which brings us back to the question at issue in *Maurice v. Judd*, the question put to Captain Preserved Fish and the sailor James Reeves: In the natural history of the whalemen, was a whale a fish? This chapter departed from the testimony of Fish and Reeves, using their different opinions on this question as an invitation to investigate what American whalemen knew about the animals they pursued across the oceans of the

---

[81] The song seems to hail from the 1840s. See Frank, "Classic American Whaling Songs," pp. 12–14.

[82] Reynolds, "Mocha Dick," p. 378.

[83] Frank, *Meditations from Steerage*, p. 30. Cooper underlines the point in his 1849 American gothic novel, *The Sea Lions*: "The word 'amphibious' is, or rather *was* well applied to many of the seamen, whalers, and sealers, who dwelt on the eastern end of Long Island, or the Vineyard, around Stonington, and perhaps we might add, in the vicinity of New Bedford. The Nantucket men had not base enough, in the way of terra firma, to come properly within the category." James Fenimore Cooper, *The Sea Lions* (Amsterdam: Fredonia, 2002 [1849]), p. 57. See also p. 372, which features a *Maurice v. Judd*-inspired joke about Mitchill's crazy new ideas about "amphibby" whales.

nineteenth century. Using published works and manuscript sources, I have sought to demonstrate that—for all the challenges that confront any effort to recover and generalize about vernacular conceptions of nature and natural order—it is possible to sketch the contours of a whaleman's natural history in this period.[84] While the interplay between the worlds of bookish cetology and the perspective of the "practical whaler" cannot be overlooked (indeed the two overlapping domains were linked, as I have shown, in cycles of reciprocal commentary, annotation, and emendation), attentive readings of whaleship logbooks can provide valuable insights into how whalemen understood and approached their prey. I have acknowledged that it is impossible to formulate an adequate idea of the whaleman's whale without appreciation of the fundamentally *fungible* nature of these creatures in the whaling industry: smooth conversions between whales, barrels of oil, and American dollars were at the heart of the enterprise, and this meant that a whaleman could look at the surface of the Pacific and see cash breaching, just as he could look at kegs of oil offloaded in Sag Harbor and see an angry sperm whale on the Brazil Banks. In fact, his intimacy with the means (and costs) of those transmutations meant that he might even see blood as the hidden coin in this infernal economy: as Melville famously put it, "For God's sake, be economical with your lamps and candles! not a gallon you burn, but at least one drop of man's blood was spilled for it."[85] And such calculus appears in the logbooks themselves. For instance, the crew of the *Parnasso* off Callao in November of 1822 had to make a tough decision: whether they should take the captain into the harbor for medical attention or finish trying out the sperm whale that had crushed his face with a swipe of its tail, leaving him "scarcely to retain the appearance of anything human"; they elected to do the former and eventually cut the valuable carcass adrift, but only after taking counsel on the question.[86]

But I have also argued that it would be wrong to imagine that this literally "reductive" view of the whales, however pervasive, rendered the animals themselves invisible or indistinguishable in the whaleman's eye. Whalemen paid close attention to whales, and this included—but also went beyond—the kinds of knowledge to be expected of any successful hunters pursuing large and complex organisms in a recalcitrant environment, a project that demanded extensive familiarity with the habits, dis-

---

[84] My theme in these pages has been a version of what Richard White has called "knowing nature through labor." See Richard White, *The Organic Machine* (New York: Hill and Wang, 1995).

[85] Melville, *Moby-Dick,* chapter 45.

[86] Kendall Log #470, p. 101, collection of the New Bedford Whaling Museum.

tribution, character, senses, and physical being of the quarry.[87] I have offered the idea of a "superficial natural history" to capture what seem to me the two most distinctive and significant dimensions of the whale-men's knowledge of cetaceans. First, these men knew the anatomy of whales in great detail, and in their entire bulky extent—but only to a depth of a few inches, encompassing the whole of the "blanket" or blub-ber layer that was the primary object of the voyage. Where these exter-nal superficies were concerned, whalemen had a detailed technical ter-minology, and an elaborated "surgical" craft for the speedy and efficient butchery of these large bodies. A vernacular anatomical tradition of some significance lies in these facts, and further work needs to be done to determine how this tradition may have drawn on and contributed to mainstream investigations of zoological anatomy (which had very different—and fundamentally deep internal—preoccupations at the time). Second, this notion of a superficial natural history can be ex-tended to include an intimate knowledge of the superficies of the ocean itself. Just as beachcombing Victorian naturalists played an essential role in making the strand a significant site for the investigation of nature over the course of the nineteenth century, whalemen produced an idiosyn-cratic natural history that addressed itself with detail and precision to the surface of the sea, and preoccupied itself with training the eye to identify, track, and depict what manifested itself at this extensive bound-ary between the worlds of water and air.[88]

But did whalemen think, in the end, that these creatures they at-tended to with such unforgiving care were, in fact, "fish"? What do the logbooks have to say on this specific problem? Even a cursory review of nineteenth-century whaling manuscripts makes it clear that Captain Fish's conviction (that whales were not fish at all) would have set him at odds with the majority of his shipmates, and not only in the 1810s and 1820s, but right through to the end of the Yankee "fishery" in the late

[87] There is a considerable literature, particularly in anthropology, on the animal-knowledge of hunters. Historically oriented contributions include: John M. Mackenzie, *The Empire of Na-ture: Hunting, Conservation, and British Imperialism* (Manchester: Manchester University Press, 1988); Matt Cartmill, *A View to a Death in the Morning: Hunting and Nature through History* (Cam-bridge MA: Harvard University Press, 1993); and Nigel Rothfels, *Savages and Beasts: The Birth of the Modern Zoo* (Baltimore: Johns Hopkins University Press, 2002), particularly chapter 2.

[88] On the beach as an increasingly important locus for Victorian natural history, see "Ex-ploring the Fringes," chapter 6 of David Elliston Allen, *The Naturalist in Britain: A Social His-tory* (Princeton: Princeton University Press, 1976). Also of interest: Joel W. Hedgpeth, "*De Mirabili Maris:* Thoughts on the Flowering of Seashore Books," *Proceedings of the Royal Soci-ety of Edinburgh* (B) 72, no. 8 (1971–1972), pp. 107–114.

1860s.[89] "Our oil from the 2 fish turned out 23 bbls" appears as the log record for the taking of two small sperms on the 3rd of May, 1821 by one New Bedford ship, strongly suggesting that it was not merely the exigencies of rhyme that dictated the vocabulary in a bit of whaler's doggerel during the Gold Rush of the 1850s:

> But we have other wishes
> To make our purses heavy
> And try to catch big fishes
> So save your californgnie . . . [90]

In fact, there are hundreds of such allusions in whaling logbooks, where one stumbles continuously across asides like this one, from 1845: "saw a few hump + finback whales but we wish for other kind of fish than these."[91] A sighting of "sulpher bottoms" (the name generally used for blue whales, which were sometimes, though not always, distinguished from what are now called finbacks or simply fin whales) occasioned similar disappointment that the wrong kind of "fish" was on the horizon.[92]

More remarkable, however, is that it emerges from these same sources that the term "whale" itself was frequently used by whalemen not as the most general term for a large spouting fish with a horizontal tail, but rather as a *specific* designation for a commercially *huntable* cete, most often a right or a sperm. For instance, a disgruntled entry in a log from the early 1840s mentions a near miss: "saw a spout—proved to be fin Backs—no whales, all gone to pot." And a similar report on a slow day in June 1844, off Peru, reads: "this day light wind; saw blackfish and finbacks—nearly discouraged in not seeing any whale."[93] In the same vein, Helen C. Allen, keeping her journal as she accompanied her husband in

---

[89] Which is not to say that the captain would not have found some partisans for his view as well. See, for instance, Captain Post's spirited contention that whales "have many analogies with larger land animals" and were improperly designated fish by some naturalists. Post's view (not entirely unlike that of Mitchill, and even Cuvier) was that whales "seem to form a sort of intermediate and connecting link between absolute beasts and their more near submarine neighbors." Post was writing sometime between 1838 and 1849. See Maury, *Explanation and Sailing Directions,* where Post's views are cited in full.

[90] For "Our oil . . .": Old Dartmouth Log #470, 3 May 1821. For the poem: Kendall Log #364, p. 304. Both volumes are held in the collection of the New Bedford Whaling Museum.

[91] Kendall Log #478, 20 May 1845, collection of the New Bedford Whaling Museum.

[92] Kendall Log #411, pp. 21, 56, 72, 108, 118, collection of the New Bedford Whaling Museum.

[93] Kendall Log #213, p. 53, collection of the New Bedford Whaling Museum. Punctuation added for clarity in both these quotations. It is worth noting the pun on "gone to pot," the joke being that the whales have literally gone (in)to (the) pots.

the *Merlin* in 1869, fell into despair in the middle of the Indian Ocean after many weeks without a take, writing: "Plenty of 'finbacks' + 'sulphur bottoms' but NO whales!"[94] In the context of these entries and others like them, it appears that not only was there a percolating question through much of the nineteenth century as to whether all whales were fish, there was, on the whaleships of the period, a question as to whether all whales were even, really, "whales."[95] Nor was this the end of the peculiar usages of the whalemen (and whale-women). We are indebted to Herman Melville for preserving a choice sample of cetological argot as the vale-diction to chapter 87, "The Grand Armada," of *Moby-Dick:* describing the mopping-up operations after an encounter with a huge "herd" of sperms off the western coast of Java, Ishmael laments the small haul the boats have managed to take from so vast an aggregation of prey (they have struck whales right and left, but ended up with only a single kill); the whole affair, he notes, is "illustrative of that sagacious saying in the Fishery,—the more whales the less fish," by which is clearly meant, "the more whales you see, the fewer whales you actually catch."[96] All of which amounts to this: for whalemen, a whale was a whale until they made it a *fish.*

Parenthetically reflecting his awareness that the cetes were, in the end, infuriatingly slippery creatures, the captain of the *Arab* hedged his bets tellingly when he wrote to his wife on the 3rd of February, 1851 about his bad luck off the southern coast of Ceylon:

[94]Kendall Log #401, 27 September 1869, collection of the New Bedford Whaling Museum; emphasis in original.

[95]This usage, curious to modern readers, probably hails from the Anglo-continental traditions in which the "true whale" (or "great whale" or "common whale" or just "Whale") referred either to what we now call a bowhead whale (*Balaena mysticetus* [Linnaeus, 1758]), or to the similar-looking creature now known as a right whale (genus *Eubalaena*), while the sperm was known as a *cachalot,* and the other more common cetaceans similarly had specific monikers. For a sense of this (and an interesting period example of a learned whaleman's rejection of late eighteenth- and early nineteenth-century naturalists' efforts to distinguish between the right and the bowhead, which Linnaeus had treated as varieties of the same species but subsequent authors divided up differently), see Scoresby, *An Account of the Arctic Regions,* vol. 1, pp. 448–478 (denying the existence of the creature depicted by Lacépède, which probably corresponds to what we would now call a right whale), and vol. 2, pp. 16–19 (arguing that the early modern whaling industry in the Bay of Biscay must have pursued small baleen whales of some sort, since *Balaena mysticetus* was unknown in the area; the industry in fact pursued the North Atlantic right whale, *Eubalaena glacialis* [Müller, 1776], which was, by Scoresby's time, hardly ever seen in European waters—which helps explain his confident dismissal of its existence). For the purpose of understanding U.S. usage, what is necessary to recall is that there was a well-established English precedent for calling only the main commercial species "the whale," and that some Americans seem to have adopted that usage, even as the American industry focused primarily on a different species (i.e., sperms).

[96]Melville, *Moby-Dick,* chapter 87.

[S]till we could not strike the whale: this is the second that has served us the same luck. I have had a chance at both, and a whale that I cannot strike is a knowing fish, (or animal).[97]

Here is a coda to the whalemen's whale (though whether it is a *horizontal* or a *vertical* coda is by no means clear): On the 29th of March, 1830, the distinguished and wealthy whaling agent Charles W. Morgan took the podium at a gathering of the New Bedford Lyceum, a forum of the worthy and well-to-do of a city built on blubber. Reading from twenty-nine pages of handwritten loose-leaf, he welcomed the assembly, and introduced his topic for the evening:

> I propose for my quota of our weekly entertainment, a subject in which the people of New Bedford are all more or less directly interested, and perhaps [on] which therefore all ought to have some general information sufficient to enable us to answer the many inquiries that are constantly made by the rest of the world, to whom generally it is a terra incognita or unknown ground— *My Subject thus Shall be the whale, a sketch of the natural history of the whale.*[98]

The Philadelphia-born Morgan had risen to prominence in the New Bedford community as a partner in the merchant shipping firm of William Rotch and Samuel Rodman in the late teens. By the 1820s he had struck out on his own as an independent financier in the whaling industry, and he held stakes in (or owned outright) a number of vessels sailing from southern Massachusetts on multi-year voyages for sperm

[97] Kendall Log #255, 3 February 1851, collection of the New Bedford Whaling Museum; emphasis in original.

[98] Emphasis mine. I am quoting from Charles W. Morgan, "Address before the New Bedford Lyceum," Charles Waln Morgan Papers, 1817–1866, box 7, series Y, pp. 3 and 17, collection of the New Bedford Whaling Museum. Apparently this document, a 27-page holograph manuscript (with several additional leaves of tipped-in insertions), has never received scholarly attention, but it is a remarkable text, given that it offers a detailed answer to the question of what "practical men" knew about the natural history of the cetaceans. Significantly, it is misdated in the New Bedford Whaling Museum's library catalogue: the lecture was originally presented in 1830, and then again in 1837 with emendations and revisions. This is interesting because it means that it was originally composed before (but revised subsequent to) the publication of Beale's *A Few Observations on the Natural History of the Sperm Whale*. I am currently at work on an edited version of the Morgan manuscript, which will discuss some of Morgan's revisions and make the text more widely available. See Dan Bouk and D. Graham Burnett, "Knowledge of Leviathan: Charles W. Morgan Anatomizes His Whale," forthcoming in the *Journal of the Early Republic*.

and right whales; he would eventually open candleworks in New Bedford, and his civic activities and expanding network of partnerships would help him win a valuable contract to supply U.S. lighthouses with costly and bright-burning spermaceti.

As a mere whaling agent (who had never actually been on a whaling voyage, though he had witnessed shore whaling operations, and watched the cutting-in and trying out of a whale in Nantucket) Morgan thus had to admit that there were those in his audience who were perhaps better qualified to lecture on his chosen topic:

> I am aware there are many who hear me, who are much more familiar with this interesting monster of ocean than I am, or indeed ever wish to be,—whose eyes have often beheld their rolling huge frame amidst the waves of the Pacific, & whose hands have fearlessly arrested his mighty progress, and forced the monarch of the deep to yield up his life & his light to the bravery and ingenuity of such an atom as man appears by comparison.

But at the same time, Morgan asserted, "there are also no doubt others, who are not so well informed, to whom some things I have read and heard may not be uninteresting."

The lecture that followed collated Morgan's conversations (with the whaling captains and sailors he knew through his work as an agent, outfitter, and speculator in the industry) with redactions from his readings of Scoresby's *Account of the Arctic Regions* (which helpfully supplemented the section of the presentation on the *Balaena mysticetus;* New Bedford whalemen were much more expert where sperm whales were concerned). One by one Morgan took up nearly all the issues I have touched on in this chapter: the actual timidity of most whales, which, he noted, for all their fearsome size, mostly endeavored to flee their pursuers; the different species recognized by the whalemen and the physical characteristics, food, and value of each; the habits and behavior of the different "tribes," both when left to themselves and when under attack; and finally their size, distribution, and prospects for extinction at some point in the future. Throughout, Morgan kept a watchful eye for discrepancies between the teachings of his whaleman-informants and the books on his shelf, for instance, where parental succor was concerned: "By some naturalists the whale is said to cary [*sic*] its young one on its back supported by a fin, but practical men do not seem to confirm this."[99] At every turn he was

---

[99] Charles W. Morgan, "Address before the New Bedford Lyceum," Charles Waln Morgan Papers, 1817–1866, box 7, series Y, p. 24, collection of the New Bedford Whaling Museum.

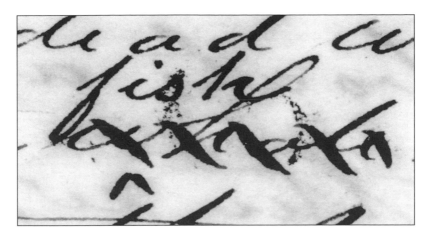

FIGURE 16. Is a Whale a Fish? Charles W. Morgan has a second thought (1830). Courtesy of the New Bedford Whaling Museum.

carefully deferential to the general expertise in the room, closing with an expression of hope that "some of our practical men may be induced at a future time to give us the results of their experiences which could not fail to form a very useful & instructive lecture."[100]

Indeed, it was apparently out of such deference that Morgan touched delicately on the matter of where whales belonged in the order of nature:

> Whales are not classed by Zoologists with the other inhabitants of the deep, but with Quadrupeds or a class distinguished by having lungs fitted for breathing the external air, organs for suckling their young and as the name implies, four supports resembling hands or feet, but as this last characteristic will not apply to all, particularly the whale, the name of this class has been properly changed by Linnaeus to Mammalia, from the organ for suckling their young which is common to the whole class.[101]

But, he acknowledged, reasonable people could disagree about the virtues of this place for the whale, and Morgan diplomatically set the knotty matter aside, recalling a certain New York legal fracas still fresh in New Bedford's collective memory in 1830:

> From the above arrangement it has been doubted whether a whale was really a fish, which we shall not argue at present,

[100] Ibid., p. 29.
[101] Ibid., p. 7.

though we all remember the noise such a discussion made a few years since.[102]

Putting aside the notorious case of *Maurice v. Judd,* what did Morgan himself think? He was cagey to the end, and perhaps not himself decided. After all, a close look at his manuscript reveals a holographic substitution that speaks volumes about the durable taxonomic ambivalence of the cetaceans: narrating the capture of a polar whale at the very end of his talk, Morgan appears to have had trouble making up his mind, crossing out "whale" and writing in "fish" just above the line [Figure 16].

---

[102] Ibid., p. 8.

FIVE

Men of Affairs

THE WHALE IN THE SWAMP

It would appear that Captain Preserved Fish was not in attendance that day at the New Bedford Lyceum to hear Charles W. Morgan hold forth on "the natural history of the whale," since it seems likely that, had he been, Morgan would have made some gesture at the Captain's involvement in the still notorious trial. But his absence is not really surprising. Though Fish came originally from New Bedford, sailed from that port as a whaleman, and retained family and business connections there, he had long since made New York his primary place of residence. In fact, by the time of *Maurice v. Judd* in 1818, Fish had already put the quarter-deck well behind him, and had been for three years in a merchant partnership with his cousin, the savvy Quaker factor Joseph Grinnell, who would go on to become a New Bedford grandee. Their fast-growing firm, Fish and Grinnell, did a profitable business in, among other things, whale oil.

Which helps explain the effort on the part of the plaintiff's lawyers to get the judge to discount much of Fish's testimony in *Maurice v. Judd*, since, in the words of Anthon, Fish (and a number of other witnesses) had "an interest opposed to the full operation of the law."[1] That is to say, as an oil merchant with dozens, even hundreds, of barrels of whale oil on hand, Captain Fish was no stranger to the "fish oil inspector" James Maurice, and there was no love lost between them. Indeed, Fish admitted on the stand that he had lately "submitted to have his oils inspected, though much dissatisfied with the law, because he did not like to stand alone in resisting."[2] In other words, he had been afraid of getting fined by Mr. Maurice, now the plaintiff, for denying him the right to inspect the whale oil in the warehouse of Fish and Grinnell. That Fish construed the term "fish oil" narrowly (and invoked his whaling experience to lecture the court on the un-fishy nature of whales) takes on a rather different character when viewed in the counting-house light of his business dealings.

[1] *IWF*, p. 21.
[2] Ibid., p. 45.

Captain Fish, both whaleman and oil man, thus offers a convenient pivot as we turn to the fourth and final group of citizens to be considered in this study. Following Sampson's parsing of the relevant parties to the question ("Doctor, you have mentioned three classes of men, fishermen, artizans, and men of science. There is a much larger class, those who neither fish, manufacture, nor philosophize; have you ever thought it worth while to pay attention to their opinion?") this book has taken up knowledge of whales first among ordinary New Yorkers, then among the community of natural history savants, and, in the previous chapter, among practical whalemen. We are left, then, with the "artisans," with those who "manufacture"—in short, with what we might call the "men of affairs."[3] It will be the purpose of this chapter to situate an emerging taxonomy of marine nature with respect to several established taxonomies of labor and market, and in doing so to unfold the fractious and powerful interests that held stakes in the question of whether a whale was a fish in New York in the early nineteenth century. *Cui bono?* we must ask of systematics, just as carefully as of political machinations and legislative wrangling; in fact, a close examination of *Maurice v. Judd* suggests that an overly tidy distinction between these domains—politics on the one hand, and the classification of natural kinds on the other—is, in the end, by no means easy to maintain. For this reason, the case provides, I will argue in conclusion, a compelling example of the perennially reciprocal constitution of natural and social orders.

To open *Maurice v. Judd* along this incision, we need to remind ourselves that four elite attorneys are seldom recruited to contest a trivial question, or indeed a merely academic one. The fine at issue in the case, $75, was a paltry sum for a serious merchant, equal to a few bolts of good quality cotton cloth, or several heaping bundles of Maryland tobacco; Judd's legal fees alone would surely exceed it.[4] How then had this case come to trial in the first place? What was really up for grabs?

For Fish, Judd, and indeed for all five oilmen and merchants who took the stand in the trial, James Maurice was little better than a parasite on the healthy circulation of mercantile affairs, a grit in cycling wheels of commerce. One of these men, the oil dealer Thomas Hazard, testifying for the defense, asserted that he, too, like Fish, had submitted nearly fifty barrels of spermaceti oil for inspection over the last year "for

[3] Ibid., p. 26.
[4] Equivalencies based on the Baltimore prices current for 10 February 1819: Holland duck trading at $27.30 a bolt, and Maryland "yellow" tobacco at $24.00 a hundredweight. See *Cohen's Lottery Gazette and Register*, 10 February 1819, prices current.

fear of being sued, being threatened by the inspector with suits." And he had capitulated to Maurice's badgering despite the fact that, as he explained, "All the dealers in the article . . . think the law never meant, but was inappropriately applied to, whale oil."[5] Putting aside the inconvenience of the inspection (it was the responsibility of the dealer to make the casks accessible to the inspector, which could mean unpacking and repacking entire storehouses and vessels; and the dealer was also responsible for effacing the inspection marks from emptied barrels, under penalty of law), there was the simple issue of cost: the statute called for a fee of twenty cents to be levied by the inspector per inspected container, regardless of size. While the law stipulated that half this fee could be passed on to customers, the residual amount added up, particularly for large-scale dealers. Moreover, the surcharge inflated prices and hurt sales. At the same time, the fee structure distorted the mechanics of exchange, by placing a disproportionate burden on the small containers of retail operations.

Set in this context it becomes clear that the facts at issue in *Maurice v. Judd* represent not a minor commercial transaction gone awry, but rather a formal test case, backed by a host of powerful city merchants, and crafted to establish (more precisely, aimed to delimit) the scope of the recently passed state statute "authorizing the appointment of guagers [*sic*] and inspectors of fish oils," which had gone on the books in late March of 1818. Samuel Judd—as the most visible purveyor of whale oils in the city—had taken the lead on the matter, but the verdict would affect dozens of chandlers, brokers, and merchants dealing in industrial and domestic commodities.[6] Since James Maurice (himself a former dealer in oils, as it happened) had taken up his commission earlier that year, he had made it his business to interpret the term "fish oil" in the broadest possible sense, and had thus made the rounds to each of his former partners and rivals, performing (with his assistant) the most thorough of services to the state, and exacting fees for an unyieldingly complete inspection of all the "marine oils" on everyone's premises. As far as his erstwhile fellow-dealers were concerned, this represented an unjustified and mercenary extension of his mandate, since in their view it was common knowledge that "whale oil" and "fish oil" were, as the dealer John Russell put it from the stand, "distinct articles of commerce," and that the statute had been aimed at the latter only.[7] Moreover, if one

[5] *IWF,* p. 21.

[6] On Judd's visibility: see his numerous and regular advertisements in the New York papers from this period.

[7] *IWF,* p. 15.

wanted to speak precisely in the language of the market, "whale oil" and "spermaceti" were further to be distinguished, the latter being derived exclusively from sperm whales, and—as the most valuable and chemically complex of all these substances—subject to a still more discriminating nomenclature ("winter strained," "summer pressed," etc.) depending on where in the animal's body the product had come from and what further refinements it had undergone. The commodity lists published in the city's papers generally cited the closing prices of these different oils—"whale," "sperm (winter)," "sperm (summer)"—separately, and, according to the merchants who gave testimony in *Maurice v. Judd,* these were not to be confused with "fish oil," which was yet another product.

While in the witness box Captain Fish laid out the basic nomenclature of the market by explaining that Fish and Grinnell sold "spermaceti, whale, and fish oils" and "these are the names by which the oils are known in commerce . . . [s]o much so that no person who wanted whale oil, ever made an application for fish oil." That these distinctions were "the universal understanding of all merchants" could be affirmed, he noted, by a look at the "prices current" cited on any exchange.[8] The oil merchant Thomas Hazard made the same point, adding still other categories:

> The oils take their names from the animals that yield them. . . .
> One particular kind of oil is called fish oil, which comes from
> the liver of fishes, and is therefore sometimes called liver oil.
> There is, also, elephant oil, spermaceti oil, and whale oil. . . . All
> Merchants acknowledge these distinctions.[9]

In fact, as Hazard put it:

> [T]he distinction between meal and flour is not more known
> than that between whale and fish oil, or between wheat and
> buckwheat. If anyone were to order fish oil, and I were to an-
> swer by sending him whale oil, it would be the same as if he
> had ordered sugar and I had sent molasses.[10]

He confirmed the seriousness of the difference with the seal of a salesman: he had himself passed up actual deals because the products were not interchangeable, and he could cite specific occasions on which he had sent a cash-carrying customer away empty-handed who had come for "fish oil" because there was none in stock, though at that very same mo-

[8]Ibid., p. 17.
[9]Ibid., p. 20.
[10]Ibid.

ment Hazard's warehouse brimmed with whale oil to spare. Another merchant, taking the stand, told much the same story.[11] If money could speak, here it was offering eloquent testimony on a palpable taxonomic distinction.

The assertion of Maurice and the plaintiff's-side witnesses—that "fish oil" was the general term and included all the oils of sea creatures— thus did not conform, according to Hazard and his fellow merchants, to the usages of the market. Fish oil was fish oil. It was the oil derived from the livers of cod and other fish (hence the synonym "liver oil"), and if someone came into the shop and requested "fish oil," giving them whale oil, or porpoise oil, or elephant oil, or, of all things, *spermaceti oil* (generally double the price of good common whale oil, which traded at about the same price as standard grade liver oil) was unthinkable.[12]

Moreover, from the merchants' perspective, this was a distinction that had nothing to do with whether whales were fish. When queried on this matter, Hazard shrugged it off: he had been "lately told by a learned friend, that a whale was not a fish," but this was news to him, and he didn't really care, since it seemed to him that this had nothing to do with the law and its extortionate misapplication.[13] On the subject of the cetacean's internal anatomy he had even less to offer: Did the whales too have a liver? He had no idea. He had never seen whale liver oil come to market, but that was all he could say. Similarly, the oil dealer John Russell, who had fifteen years in the trade, declared himself agnostic on the niceties of book taxonomy. He had once seen a whale "along the side of the wharf" (in all likelihood that smelly carcass moored in the East River back in 1804), but said he did not know "whether it is, properly speaking, a fish or not," nor did he have any idea how many kinds of whales were recognized. He knew even less about "sea elephants" (i.e., elephant seals), from which "elephant oil" was derived. "Do you understand it to be

---

[11] See the comments by Comstock, *IWF,* p. 47.

[12] In early 1819 the best spermaceti sold for a little over a dollar a gallon, and common whale oil traded at about $.55 for the same volume. Liver oil was generally sold by the "barrel," and the U.S. barrel in principle held 31.5 gallons, but the barrels in which liver oil sold were known as "nominal barrels" which could contain from 26 to 30 gallons and thus held "about ten per cent less than a lawful barrel" (New York [State] Legislature, *Journal of the Assembly of the State of New York,* p. 263). Hence, liver oil at $18 a barrel worked out to cost very nearly the same price per gallon ($.57–$.69) as common whale oil (which had seen a price drop in late 1818; see the *Commercial Advertiser,* 2 January 1819, citing the London prices current for November 1818). For comparatives I have used several sources, including the wholesale prices current listed in the *Spectator* for 29 December 1818, and the Baltimore prices current from *Cohen's Lottery Gazette and Register,* 10 February 1819.

[13] *IWF,* p. 21.

accounted as a whale or a fish?" Anthon inquired. "I do not know the nature of it," Russell answered. "Is the oil of a whale the oil of a fish?" Anthon probed, slyly. "I cannot say," came the reply.

From the perspective of the defendant and his supporters, then, a whale might be a fish (and thus the oil of a "whale" might well be the oil of a "fish"), but this did not by any means make "whale oil" (or "spermaceti") into "fish oil." As Judd's attorney, William Price, put it in his closing arguments to the jury (building an argument in the alternative in case Dr. Mitchill's zoology lesson failed to carry the day):

> Should you not be able to agree with naturalists in one branch
> of the evidence in this cause, and decide that a whale is, in com-
> mon acceptation of the community, considered to be a fish,
> there remains still a very important question, on the just deci-
> sion of which the defendant will, I think, be entitled to your
> verdict. Is whale oil bought and sold and consumed under the
> appellation of fish oil?[14]

Price's answer was no. Or, as Price's co-counsel, General Bogardus, put it, "The position taken on the other side, that if a whale be a fish, whale oil is of course to be called fish oil, is not true."[15]

Just as the taxonomy of the market—different prices, different entries in the commodity-exchange broadsheets, different customers—could be cited as evidence for the distinction, the defense leaned heavily on what might be called the taxonomy of craft, meaning the classifications that emerged from the differing processes of manufacture. All the ordinary "whale oils" were, looked at this way, essentially tallows, which is to say that they were rendered animal fats, produced by a process of "trying" raw fat over fire.[16] Common oil from a whale was thus very much like the grease from a cow or pig, putting aside the nettlesome issue of whether a creature with no legs counted as a quadruped. At the spigot end of the copper kettles, whales and cattle looked more similar than different. Not so with what the merchants wanted to call "fish oil." This substance, as one of the oil men explained,

[14]Ibid., p. 53.

[15]Ibid., p. 55.

[16]I say "ordinary" whale oils, since as it happens the purest *spermaceti* oil (that derived directly from the "headmatter") did not need to be tried out, but was instead subjected to several different filtering operations. Similarly, porpoise jaw oil (the finest lubricant then known—used by watch- and clockmakers) was never boiled.

> Comes from cod, haddock, pollock, sharks, mackarel, and all
> kinds of fish. . . . The livers are taken out and thrown together
> into a barrel, where they melt into oil. There is always a sedi-
> ment on this; it is not boiled as the whale is, but merely left to
> the operation of the sun, so that there is some water and mucilage
> in all fish oil, and pieces of membrane and sheds of fibres.[17]

Indeed, there was so much blood and scum in this foetid leeching that it could not generally be "tried" for the purposes of purification, since it was an unsteady emulsion that spattered badly at high temperatures and easily boiled over.[18] By contrast, as another merchant explained, "You cannot mix water with whale oil no more than fire," because of the different process of its extraction:[19]

> The oil of whales and porpoises is boiled intensely on board the
> vessels when it is taken. It is rendered so hot that a drop of mois-
> ture will fly off; if you spit on it, it will fly as from hot iron, and
> if you throw any water it will boil off it, and run over the deck.[20]

These manufacturing details were essential because they went to the heart of what was understood to be the purpose of the legislation at issue: to guard against adulterated and/or unacceptably impure "fish oil" coming to market. What was the "fish oil" that had long been the target of customer complaints? "Liver oil," according to the merchants, for the simple reason that it was by its nature a sloppy substance, of uneven quality, and thus very liable to prompt dissatisfaction among purchasers. Such troubles, they asserted, were more or less unknown with expensive household illuminating oil like spermaceti, or the lubricant and lighting oils taken from porpoises and the ordinary whales. Generally tried out by fire in the process of their manufacture, these oils were comparatively uniform, consistently "dry" (without admixtures of water), and seldom occasioned any difficulties. With common "fish oil," the stuff oozed out of the livers of various groundfish, things were admittedly different.

The products had different recent commercial histories as well, since

---

[17] *IWF,* p. 17.

[18] For an interesting period discussion of techniques for clearing and deodorizing fish oil (using powdered chalk and slaked lime), see question number 5 in "Philosophical Questions, Answered," *The Monthly Journal, Containing Disquisitions in Natural Philosophy* 1, no. 3 (1818), pp. 57–64.

[19] *IWF,* p. 17.

[20] Ibid.

FIGURE 17. Gauging and Inspecting Whale Oil: Chase and Heaton depicted work on the wharves of a New England whaling town at mid-century (1861). Courtesy of Firestone Library, Princeton University.

the perpetual disputes over liver oil had given rise to a set of formal gauging regulations that preceded the passing of the new and disputed "act for guaging [*sic*] *and inspecting* of fish oil" in March of 1818 (under which the action against Judd was brought). Under those older statutes the duties of "gauging" were confined strictly to the establishment of the volumes of vessels, not the character of their contents. While there was inevitably a watchdog function in simple gauging—a gauger might, for example, draw attention to a barrel that had been "made purposely with heads thick toward the centre"—a gauger did not need to know anything about oil *per se,* and he passed no judgment on the quality or virtue of the material in the container. For instance, Ralph Duncan, a gauger who gave evidence in *Maurice v. Judd,* asserted that he was wholly ignorant of oils himself, having been trained as a bookkeeper (which made sense, since gauging amounted to the arithmetic of volumetric analysis).[21] Figure 17 shows gauging of barrels of whale oil at a New England port in the mid-nineteenth century, complete with a clerk (on the left), doing the math. According to Duncan, under the old New York gauging law whale oils had been universally disregarded, since those casks— all different sizes because they were custom made to fit optimally in the hold of a particular whaling ship—were always gauged immediately

[21] For problems with casks, see *IWF,* pp. 24 and 45.

upon landing at the wharves of the whaling ports; it was only with the advent of the new "inspections" provision (and Mr. James Maurice) that the whale oils had come under scrutiny in the city.[22]

The scrutiny in question involved the application of a specialized instrument, known as a "trier." Most versions consisted of a sounding rod enclosed in a sheath, something like a ramrod sliding in its gunbarrel (the gentleman in the center-background of Figure 17, standing on a cask, is using a similar device—a siphon trier).[23] By slipping the paired rods into the bunghole and then withdrawing the inner rod, the inspector could sample the deep matter in the cask cleanly, and avoid the pitfall of ordinary "dipstick" sounding: any simple rod would come up nice and oily, since whatever oil there was in a container floated on the sludge and water.[24] Indeed, before the law came into effect, barrels of liver oil were known to have come to market containing as little as two or three gallons of actual oil, the remaining thirty or so being sediment and dirty fish water.[25] Because the new law authorized the inspector to mark each barrel not only with the gauged volume of the cask, but also with his assessment of the "split" of oil, water, and "soot," such fraudulent (or, putting aside the issue of intent and malice, deceptively inferior) parcels of liver oil had grown increasingly rare since March.

In this sense, the law was working, and that was certainly the position of the man who drafted it, who took the stand in *Maurice v. Judd* to defend his legal innovation against what he saw as a cabal of self-interested oil-men and zoological metaphysicians. Gideon Lee was sworn as a witness for the plaintiff. Asked to state his profession, Lee demurred: "I hardly know by what name to call it. I deal in hides, skins, leather, oil, &c.; am interested in several tanneries, and carry on the currying busi-

[22] The term "cask" (unlike "barrel") did not generally denote a measure (though in dry goods it sometimes could), but rather an object—the vessel made by a cooper to contain liquids. In the whaling industry casks were made en route by an on-board cooper, who assembled different-sized casks in such a way as to maximize the stowage of his vessel. There are a number of "cask diagrams" that survive for different whaling ships, and they lay out both how the casks were to be packed in the hold, and the different sizes that they should be for this purpose (casks were measured by diameter of their heads). For a discussion: Mark Howard, "Coopers and Casks in the Whaling Trade, 1800–1850," *Mariner's Mirror* 82, no. 4 (November 1996), pp. 436–450.

[23] Several whaling museums hold examples of gauging and inspecting instruments. The former look like large calipers (for measuring the external dimension of the casks; for an example look in the foreground of Figure 17); there were a number of different kinds of "trier" used over the course of the nineteenth century, including the "thief" (a kind of slim siphon) and oversized glass pipettes that made it possible to see the contents (for the purpose of assessing color and quality).

[24] On the trier: *IWF*, p. 47.

[25] Ibid., p. 41.

ness."[26] He was being modest. The forty-year-old Massachusetts-born Lee had arrived in New York City about a decade earlier, and traced a meteoric ascendancy to a position of preeminence in one of the city's oldest, toughest, and most tightly knit communities.[27] Known as the "Swamp" (with equal parts affection and repugnance), the southeastern blocks of the city centered on Jacob and Ferry streets [see Plate 4, just east of City Hall] had been the (stinking) locus of the tanning and leather currying industry for more than thirty years. At the very crossroads of this neighborhood Lee had erected in 1808 the building he called "Fort Lee," which soon became the largest leather emporium in the area, and the foundation for what would become a century-long dynasty in the leather trades in the United States. As Lucius Ellsworth puts it in his economic history of the New York State tanning industry, "Gideon Lee dominated the New York leather industry throughout most of the first four decades of the nineteenth century." And New York leather was king in the United States. The turn of the century chronicler of the Swamp, Frank Norcross, asserted that "the name of Gideon Lee was perhaps more widely known than that of any other merchant of his time," and Norcross noted that Lee wore his reputation as a leatherman with pride even as he outgrew the Swamp precincts, rising to the mayorship of the city of New York, 1833–1834, and eventually to national office.[28]

For Lee, leather was more than a smelly trade and a cobbler's commodity. It was a philosophy, a way of life, even the measure of civilization itself: Lee was known to invoke "the great metaphysician Locke" in defense of the proposition that tanning outstripped all the "merely mechanical" trades in difficulty and subtlety, since the tanner dealt not only with "number, figure, color, and dimension" (those readily accessible, even trivially obvious, features of the world), but also with the "texture

[26] Ibid., p. 40.

[27] Lee is introduced in Sean Wilentz, *Chants Democratic: New York City and the Rise of the American Working Class, 1788–1850* (New York: Oxford University Press, 1984), pp. 37, 41 (note 37). See also: "Sketch of the Life and Character of the Late Gideon Lee," *Hunt's Merchants' Magazine* 8 (1843), pp. 57–64; and the citations below. There are Lee papers both at the New-York Historical Society (mostly from his term as mayor) and in the New York State Archives in Albany (which were more relevant to this project). Ellsworth cites some materials that were at Rutgers in the early 1970s, but these I have not seen.

[28] Lucius F. Ellsworth, *Craft to National Industry in the Nineteenth Century: A Case Study of the Transformation of the New York State Tanning Industry* (New York: Arno Press, 1975), p. 62; Frank W. Norcross, *A History of the New York Swamp* (New York: Chiswick Press, 1901), p. 51. Within this latter text Norcross refers to the "present century" in ways that make it clear that he is talking about the nineteenth, so I have chosen to situate his remark then, despite the publication date of the (self-published) book.

of substances," that most ineffable aspect of matter, which "is hardly as-certained or comprehended by the most extraordinary minds." For this reason, Lee asserted, tanners had to be sober and gifted, possessed of "the keenest perceptions, the most vigorous mind, and the soundest judgement." Where many tanners were found, there lay much sagacity and virtue: as the Swamp prince proudly noted (here paraphrasing Adam Smith as part of an effort to "re-brand" just about the dirtiest and most stinking craft known to humanity), "it is characteristic with savage nations to export their raw hides, and neither to manufacture nor use much leather; while civilized nations import largely of raw hides, and manufacture and consume large quantities of leather." According to this idiosyncratic measure (leather production = degree of civilization), by the late 1830s, thanks largely to Lee himself, New York rivaled London at the pinnacle of cultural progress on the globe.[29]

While he trained young as a shoemaker and tanner, and traveled as a leather salesman in his youth, Lee's rapid success in the trade had less to do with craft knack than with his exceptional financial acumen and cre-ativity. In his first decade working in the city Lee pioneered a set of in-novative contract arrangements that would become industry standards in the 1820s and 1830s. Most significant of these was a modification of the system that had made the older New York tanner-mogul Jacob Lorillard one of the wealthiest men in the city in the same period. Lorillard had created a state wide web-like leather empire by contracting out tanning operations for a fixed rate to small country enterprises up the Hudson Valley (which were closer to the increasingly remote sources of the most important ingredient in the process, tannin, mostly derived from hem-lock bark). Lorillard himself owned the hides, and if he could bring the resulting leather to market at a price exceeding his costs for the raw skins, the tanners' fees, and the transportation, he made money. And he did, amassing an estate worth nearly half a million dollars by the time of his death in 1838. Lorillard's scheme grew to dominance in the period of instability following the War of 1812 because it guaranteed tanners a fixed income for their work during difficult years, and the spread of his arrangement helped make tanning an even more capital-intensive enter-prise than certain practical aspects of the trade already necessitated (the long periods of time required to tan many kinds of leather, some of which could demand more than a year of costly and laborious steeping,

[29] These quotes and the relevant statistics appear in Gideon Lee, *Two Lectures on Tanning* (New York: The Eclectic Fraternity, 1838), and reflect his mature views on the industry.

meant tanners always had large up-front expenses). The young Gideon Lee transformed Lorillard's system to distribute the capital investment while retaining access to a considerable portion of the profit. Rather than buy any raw hides outright, Lee acquired them for his tanners on accounts he established for them, charging a commission for this service. Like Lorillard, he assumed responsibility for transportation, shipping the hides out, and bringing the finished leather back to market (for the tanner), where he sold the final product, again on commission. Configuring the dealings this way, Lee was really a kind of broker, but he also developed a unique fixed-price option, which permitted tanners to lock in a future price for their tanned hides (in return for a supplementary commission). Small upstate tanneries could thus sell their risk, but at a substantial discount on their potential returns. Possessing a savvy eye for the market, and willing to impose aggressive fiscal demands on his tanners (for instance, he demanded payable notes for the raw hides on delivery), Lee positioned himself to make large sums of money while tying up comparatively little principal.[30]

In the process, Lee expanded into parallel ventures. Like other Swamp business men, he extended loans to upstate tanneries in the Catskills (and later in the Adirondacks), frequently affording funds for tan yard expansions in return for shares in these establishments.[31] It was in pursuit of similar arrangements of shared ownership that Lee and several partners formed, in June of 1817, the first formal joint-stock corporation for tanning, the New York Tanning Company. At the heart of this venture was a Massachusetts leather dealer and entrepreneur, William Edwards, with whom Lee had worked before establishing himself in New York. Edwards had followed Lee to the city after the bankruptcy of a Massachusetts tanning enterprise, and with Lee's help Edwards secured a set of investors willing to back him in the creation from scratch of a new large-scale tannery that would take advantage of its high capitalization to invest in novel technologies that promised to reduce labor costs and speed up the tanning process. The financial arrangement Lee and Edwards were seeking was sufficiently novel that it demanded changes in the incorporation laws of the state of New York, changes they were able to secure through the intercession of Edwards's cousin, a judge in the state courts, who introduced a bill in the state legislature to

---

[30] These different arrangements (as well as several hybrids) are the subject of chapter 3 of Ellsworth, *Craft to National Industry.*

[31] For a remarkably detailed account of these deals, see Barbara McMartin, *Hides, Hemlocks and Adirondack History: How the Tanning Industry Influenced the Region's Growth* (Utica, NY: North Country Books, 1992), particularly chapter 3.

modify the relevant statutes.[32] When that bill passed, in April of 1817, the way was cleared for a new kind of corporate-industrial tannery, vastly larger than the hundreds of family-sized operations that dotted the state: 1,200 acres were secured in Greene County, and a tan-yard constructed with a capacity to produce 10,000 "sides" (half-hides) of sole leather each year. This massive establishment had the distinction of being the first covered tannery in the United States, and Lee sank money into warehouses along the water route to the Hudson, so that finished hides could be stored until the ice broke in the canals.[33] Six partners made initial investments of $10,000 each in the New York Tanning Company, including the Cuban merchant, Joseph Xifie, who helped secure the first five thousand raw bullock skins from Argentina.[34] The first of those skins, tanned to sole leather, came to market in New York in autumn of 1818.[35]

It is thus safe to say that Lee, like Mitchill, needed little introduction in the Mayor's Court. And this well-known man of affairs had come prepared to cross swords with the celebrated naturalist. Asked if he had been in the court earlier that day, and understood the issue at law, Lee replied that he had indeed heard the morning's testimony, and had thus discovered that he would be obliged to "contradict some worthy and respectable citizens" (here meaning above all Mitchill) on the matter of whales and fish.

As the acknowledged author of the text of the "act authorizing the appointment of guagers [sic] and inspectors of fish oils," Lee had come to court to back the inspector (perhaps even, we might say, his inspector), James Maurice, and to explain to the jury the proper scope of the legislation, legislation Lee had conceived, driven to the attention of Albany lawmakers, and minutely annotated by hand in its draft form.[36] Examined in the context of Lee's simultaneous business ventures, and his

[32] Ellsworth, *Craft to National Industry,* p. 66.

[33] See Lee's warehouse ads in the *New-York Daily Advertiser,* 1 January 1819.

[34] On the New York Tanning Co., see Norcross, *A History of the New York Swamp,* p. 52. It surprised me to learn that about a third of hides tanned in the United States came from Spanish America in this period.

[35] Norcross, *A History of the New York Swamp,* p. 52. While the New York Tanning Company was a pioneering venture, it was not, in the end, a financial success: see "Leather Manufacture," *Hunt's Merchants' Magazine* 3 (1840), pp. 141–148, which also provides an account of the later development of similar (and more lucrative) large tanneries, the most successful of which was the Prattsville Tannery (also built in Greene County) of Colonel Zadock Pratt, probably the largest tanning establishment in the world by the late 1830s, with a capacity for 50,000 "sides" a year. It is notable that Pratt started out as a tanner for Gideon Lee.

[36] I have secured from the New York State Archives at Albany a manuscript copy of the fish oil act from the Gideon Lee Papers, box 18, folder 4. The document appears to be in a

other successful recent lobbying efforts to secure legal protections for his schemes, the fish oil act at issue in *Maurice v. Judd* must be understood not as merely some minor regulatory initiative to protect jug-carrying oil customers in the Fly Market from adulterated wares, but rather as yet another element in an increasingly ambitious effort by Lee (and an emerging consortium of politically connected tanner-entrepreneurs) to push laws favorable to the commercial leather trade through the state assembly: leather workers were, after all, major consumers of fish oil, and Lee—who bought very large quantities of such oil for his own enterprises, and to supply to others—must have decided in late 1817 (shortly after the successful emendations to the state's incorporation laws for the benefit of the New York Tanning Company) that legal mechanisms were the best way to clean up a messy and unreliable dimension of the industry he intended to make more efficient and profitable.[37]

Tanners and curriers (the latter focused on the finishing of hides that had been tanned by the tanners, though curriers often also undertook the full processing and dressing of lighter skins from calf, goat, and sheep) made use of a number of oils at different stages of their crafts. Of these, liver oil from fish—sometimes also known as "currier's oil"—was the most important, serving as the main ingredient in the preparation of "oiled leather," or *chamois,* still used in clothing, aprons, washcloths, and the like in this period.[38] In addition, nearly all shoe leather for uppers

---

scribal hand, but there are marginal additions and corrections which closely match the hand of Lee himself. Under oath, Lee answered in the affirmative the following question: "Did you draft the bill under which this prosecution was brought?" (*IWF,* p. 42). It is also worth noting that James Maurice had been in a commercial partnership with one "Edwin Lee" in New York City as late as 1818. I have not been able to establish if Edwin and Gideon were related, but a link is likely. It seems reasonable to suspect that Gideon Lee helped Maurice secure his position as city gauger and inspector.

[37] See Wilentz, *Chants Democratic,* p. 401 (table 6) for evidence of the growing wealth and political clout of the New York tanners in this period. It should be noted that Edwards (and others) had been advocating such legal intervention in different elements of the business for years. See "An extract from William Edwards of Northampton to the Postmaster at that place, dated November 29, 1809" published in the *Philadelphia Repertory* 1, no. 4 (26 May 1810), pp. 31–32. Edwards there called for regulations on slaughterhouses to prevent damage to raw hides. Leather inspections, too, were a regular aspect of the business; see "Leather Manufactures," *Hunt's Merchants' Magazine,* p. 146. The more general issue of inspection and regulation of consumer products was controversial in the late teens both in the United States and in Britain. For a sense of the concern, see Friedrich Christian Accum, *A Treatise on Adulterations of Food and Culinary Poisons* (London: J. Mallett, 1820), and the discussion of this text in the *Spectator,* 1 June 1821, p. 2.

[38] There is a literature in the history of technology dealing with tanning, but much less material available on curriers. For an introduction to the processes (and an appendix of useful extracts from relevant primary sources), see Peter C. Welsh, *Tanning in the United States to 1850*

was oiled, absorbing between a quarter and a third of the weight of the tanned skins. The thicker, stiffer sole leathers, too, received a coat of oil after drying. Trade secrets shrouded the exact composition of many of these oils (which were frequently stiffened with tallow, thinned with linseed oil, or layered with wax—depending on the leather, its purpose, and the qualities of the oils themselves), but fish oils were the main ingredient in these concoctions.[39] For a major leather dealer like Lee, who might have a stake in upwards of 150,000 hides at any given moment, hundreds of barrels of fish oil (and plenty of other fats, too) would be bound up in the calculus of his affairs.[40]

(Washington, DC: Smithsonian Institution, 1964). Much of this text focuses on the Delaware Valley industry. See also Peter C. Welsh's "A Craft That Resisted Change: American Tanning Practices to 1850," *Technology and Culture* 4 (1963), pp. 299–317. The most comprehensive source, however, is the curious composite text published in 1852 as *The Arts of Tanning, Currying and Leather-Dressing* by Campbell Morfit (Philadelphia: Henry Carey Baird, 1852), which is dedicated to Zadock Pratt. This purports to be an "edited" translation of the French manual from 1833 on the same subject by Julia de Fontenelle and François Malepeyre (under whose authorship Morfit's volume sometimes appears in library catalogues), but Morfit in fact amended and altered this original text into a manual on contemporary U.S. practices. The result is more than 550 remarkably detailed pages covering the chemistry, mechanics, nomenclature, and labor practices of the craft, and giving step-by-step instructions on the dressing of everything from human skin to coach leather. On oils in general, and fish oil in particular, see pp. 433–439, 478–484, and 489–492.

[39] It should be emphasized that craft practices differed, and hides demanded particular attention depending on their raw condition (salted or dried, fresh or already rotting, thick or thin), the character of the tan (different bark types, different lengths of time), and the weather (temperature, humidity, etc.), so it is difficult to generalize about the process. The final products, too, of course, were very different, ranging from fine glove leather, to blacked harness leather, to belt leather for use in driving millworks—each type demanded particular techniques of liming, bating, washing, tanning, and dressing. There were hardly any leathers, however, that never saw oil in the course of their production.

[40] Lee testified in *Maurice v. Judd* that he had bought "as many as ten parcels a year" of fish oil (*IWF*, p. 42), primarily from the main dealer in the United States (located in Boston, since the vast majority of the cod fished in the western Atlantic were taken off Georges Banks). I have not been able to find any records of these purchases in the manuscript material surviving from Lee's establishments, nor are they broken out of the economic analysis of the industry by Ellsworth. Given that Lee alludes to the problem that his buyers could only examine "a few casks of the uppermost tier" of a parcel of oil that changed hands within the ship that brought it to market (*IWF*, p. 40), and that this was an unacceptable sample of the lot, it is reasonable to assume that these parcels were seldom fewer than a dozen barrels. Making the calculation a different way (by the weight of hides Lee handled, and the trade-manual guidelines for oil per pound of leather) I came up with an estimate of 140 barrels per year. While I think this is probably a considerable underestimate (Lee also supplied tanners with their materials, and thus surely handled oil that did not go into his own hides), it should be acknowledged that it is a very rough method, since the amounts of oil needed vary so widely by the type of leather in question and specific condition of the skins. For a sense of Lee's stock, see his advertisement in the *Mercantile Advertiser* of 12 January 1819, p. 3, which demonstrates that he had some 28,000 hides on hand (as well as more than 150 barrels of various oils).

It was thus with mounting irritation that Lee had witnessed a day's worth of sophistical dissection of fish, whales, and nomenclature. As he explained on the stand, it was true that tanners and curriers principally used "liver oil" in their trade, but, "in the absence of this, as was the case during the late war, the curriers used every kind of oil drawn from marine animals." Moreover, all such oils were known to come to market in varying states of purity, and while Lee admitted that he "had not dealt much in the different kinds of whale oil" (and was willing to acknowledge that it was generally "much less adulterated than the liver oil"), nevertheless, it was unquestionably his view that "all kinds of oil require inspection."[41]

The notion being bandied around the court—that "fish oil" meant uniquely "liver oil"—was the fiction of a clique of evasive chandlers:

> We use the general term, fish, to comprehend them all; a less general phrase is whale oil; of this there are several species, as sperm, humpback, common whale, &c. We have also sea elephant oil and porpoise oil. Another class of fish yield oil from the livers only, such are the cod, haddock, pollock, &c.; this we call liver oil; there is also dog and manhaden [*sic*] oil; sometime we use the phrase, shore oil; this is a mixture of different oils taken on our coasts; and I have heard the fishermen apply the term, ground fish, to such as grovel on the bottom, in contradistinction to those fish that rise to the surface of the water for the purpose of breathing.

To which the examining attorney replied with the essential question, "Who do you mean by we?" giving Lee the opportunity to invoke the stolid authority of a rich man in a town built on the dollar:

> I mean we, plain men, who carry on the substantial business of this great world, and however naturalists have made other distinctions and classifications for other purposes, we disregard them. In the course of our common pursuits we call all marine animals fish, and land animals quadrupeds.[42]

As for the argument of the merchants, that the prices current on the various exchanges distinguished among the marine oils, he was willing

---

[41] *IWF,* p. 41.

[42] Ibid. It is striking to watch Lee do a little twist on the meaning of "ground fish" here. The term was hardly intended by fishermen to distinguish fish-fish from spouting fish (as Lee implies by his testimony), but rather to distinguish bottom-feeding fish (cod, flounder) from top-water fish (mackerel, striped bass, etc.).

to credit that much, but he denied that the term "fish oil" ever appeared in such documents to mean, specifically, curriers' oil. The term "liver oil" was used, and frequently sat next to whale and sperm oil in such listings, but that proved nothing about how the term "fish oil" itself was used in the common language of traders. "Fish oil," he insisted, could—indeed it *did*—mean all those oils that came from the creatures of the sea. To prove as much, he dropped on the table in the court a stack of letters, invoices, bills of lading, and printed circulars from Holland, France, London, and several U.S. cities, all culled from the offices of "Fort Lee" and all purporting to establish that "fish oil" could be used to mean any marine oil.[43]

While this material evidence seemed to undercut the argument of Hazard, Comstock, Fish, and others—to wit, that the market consistently distinguished whale oil from fish oil—a crafty cross-examination backed Lee into a reluctant acknowledgment that money made its own taxonomic distinctions. Hypothetically speaking, Judd's lawyers mused, if a correspondent were to order from Mr. Lee, say, fifty barrels of "fish oil," what would the gentleman send from his stock? Lee hedged: "If he were a currier, or I was aware that he knew I seldom dealt in any other than liver oil, I presume I should send such." Which was as much as to say, was it not, that he would be very unlikely to send, for instance, *sperm* oil? To which Lee, ever the business man, replied: "I would deliver you the very cheapest kind of fish oil that I could find in market, presuming you understood your business." In other words, *not* sperm oil? "Certainly not, sperm oil generally bears the highest price of all."[44]

So it was perhaps not quite so clear that "sperm oil" was, in such a case, "fish oil," since Lee would be loathe to fill an order for "fish oil" with, say, three casks of spermaceti like those at issue in *Maurice v. Judd*. And with that, Lee was asked to stand down.

But Lee had not come to court alone. On the contrary, the roster of witnesses for the plaintiff reads like a who's who of the Swamp. Jacob Lorillard himself took the stand, a man who owned more than one hundred prime lots in Manhattan (including a fair portion of the Swamp itself) and who would pay taxes on $600,000 worth of property the next year.[45] In 1822 he would buy out Lee's New York Tanning Company, and

---

[43]Unfortunately, these documents were not preserved or closely cited in the court records, but I can confirm that in several months of archival work I have yet to find the term "fish oil" in a printed price current for the period, while entries for "liver," "whale," and "sperm" oils regularly appear.

[44]This exchange appears in *IWF*, p. 42.

[45]Norcross, *A History of the New York Swamp*, p. 61.

the two men would be fast colleagues to the end of Lorillard's career in 1834, when Lee would lead a delegation to present the retiring Swamp king with a set of commemorative silverplate. Lorillard, too, testified that he had ransacked his office in search of a document that used the term "fish oil" in the limited sense alleged by the oil merchants who had given evidence for the defendant; he asserted he could find no such thing. Indeed, returning to the matter of oil manufactures, Lorillard met the oilmen on the grounds of their taxonomy of craft: it was impossible that "fish oil" could be used exclusively as a synonym for liver oil as they alleged, since "manhaden [*sic*] oil" (frequently used by curriers) was unquestionably a fish oil—menhaden, recall, are a small sardine-like fish that swarms along the East Coast in the summer—but it was made not from the liver but rather from the whole fish, which were shoveled *en masse* into presses to squeeze out a body oil.[46] So here was a "fish oil" that was by no means a "liver oil." Clearly, then, "fish oil" was a more general term. Citing his fifteen years in the tanning business, Lorillard confirmed Lee's testimony to the letter: "fish oil embraced every species of marine oil" in the "understanding of the trade," and all those oils ought, in Lorillard's view, to be inspected.

Similarly, one of Lee's main investment partners in the New York Tanning Company, Hugh McCormick, whose Swamp offices were a stone's throw from those of Lorillard and Lee, took the stand to insist that tanners were by no means unacquainted with whale oil, since it was not infrequently mixed in with "other fish oil" to make various finishing treatments. "Fish oil means all sorts of marine oil," he informed the court, and warned that oil merchants' protestations about the cleanliness and reliability of the tried-out oils were not to be credited: he had bought a fair bit of whale oil over his twenty years in the business, and "found sediment and mixtures of worthless and unmerchantable matters in it, and water." Inspections were emphatically necessary to put an end to all the "deceptions and frauds" in the oil markets.[47]

All these opinions were echoed by the proprietor of the Ferry Street firm of Van Nostrand and Co., and by several other senior Swampers, including the eccentrically austere Quaker magnate Israel Corse, a familiar figure in the city, known by the silhouette of his Society of Friends attire and later for his habit of being picked up each morning from the marble-pillared front step of his mansion on East Broadway by the indecorous tumbril that carried the heap of the day's hides down to the

[46] *IWF,* p. 44.

[47] On McCormick and the firm of Cunningham & McCormick, see Norcross, *A History of the New York Swamp,* pp. 12–13. The quotes come from *IWF,* p. 46.

shop of Israel Corse and Son.[48] According to Corse, most dealers in New York would probably send him liver oil, were he to place an order for "fish oil" without giving any more detail, but if the dealer did not know that the order had come from a tanner, "he would be as likely to send whale as any other oil," since "fish oil, in a general sense, embraces the whole of the marine oils."[49]

By the time this parade of tanner witnesses retired, the underlying contest of *Maurice v. Judd* was impossible to miss. As John Anthon put it in his closing statement to the jury, the case was, at its core, "a struggle between the venders on the one side, and the purchasers on the other," where each side was standing up for its interests. As counsel to the plaintiff—an inspector who had been commissioned by, and effectively worked for, the "purchasers" (i.e., the tanners and their financial backers)—Anthon made it clear where he stood: those who bought oils for industrious use were merely "insisting on inspection for their protection," a protection that would help them bring quality products to market; by contrast, the venders were predictably trying to dodge such oversight for their own pecuniary gain.[50] But the contrast between dealers and consumers could cut the other way, too, since, as Judd's lawyers were quick to point out, who could be expected to have more extensive knowledge of the range, characters, and nomenclature of the marine oils, dedicated dealers (who dealt in them all, all the time) or mere tanners (who mostly stuck to [fish] liver oil, and hence were really only familiar with that one product)? William Price, summing the defendant's a-whale-is-not-a-fish case, leaned heavily on this point, and contrasted the very different horizons of tanners and merchants in a bid to establish the greater expertise of the latter:

> All the plaintiffs [*sic*] witnesses are leather dealers or curriers, who seldom if ever make use of the oil of any whale, and never, perhaps, of the spermaceti; whilst those on our side, are extensive

[48] On Corse and the shops spawned from his original establishment, see Norcross, *A History of the New York Swamp,* pp. 43–45 and passim.

[49] *IWF,* p. 45. This idea—that tanners generally got liver oil because everyone knew they were tanners, whereas in an anonymous market, an order for fish oil might be filled with any marine oil—comes up several times in the transcript, and offers an interesting glimpse of a tacit distinction between what we might call "the market in principle" (an anonymous exchange of goods for money) and "the market in practice" (where all of the major participants were bound into social and kinship networks that shaped the exchanges). See the clearest statement by the witness Thomas Brooks (who may have been a member of the swamp tanning firm of Henry Brooks): "Different people have different ideas," he noted, but if he were to get an order from "an entire stranger for fish oil, he might very fairly send him whale oil." *IWF,* p. 46.

[50] *IWF,* p. 59.

dealers in whale oil, and, necessarily, know every fact connected
with their own branch of trade.[51]

It was an argument that Sampson sagely turned back on itself: the more
the oil men trumpeted the magnitude of their dealings in all the various
marine oils, the more they adverted to their larger financial interest in
the outcome of the case, and in doing so cast the lengthening shadow of
self-interest over their casuistically impassioned distinctions among the
denizens of the sea.[52]

By taking up the activities and preoccupations of men like Israel Corse,
Jacob Lorillard, Gideon Lee, and Thomas Hazard, this chapter has revealed that the case of *Maurice v. Judd* represented above all a showdown
between two powerful groups of politically active and wealthy figures
in the city of New York in the early nineteenth century: a clique of oil
merchants and chandlers on the one hand, and a consortium of tanner-
financiers on the other. Placing the whale in the thick of Swamp politics
in the significant years 1817–1818 has demonstrated that the boundaries
of natural history would have to be drawn not merely on the blank pages
of student notebooks at the College of Physicians and Surgeons, but
also across the minutely inscribed leaves of ledger-books and legal
codes—indeed, across a densely populated landscape of labor and exchange, where a crosshatching of closely guarded boundary lines had already been settled by custom, and each move exacted differential costs.
Within these contesting communities the question of whether whale oil
was fish oil did not reduce cleanly to the question of whether whales
were fish. What students of comparative anatomy or observant whalemen might offer on this matter—yea or nay—could not supersede a set
of larger considerations about the place of these different creatures in
shelved inventories and open-lot tryworks. I have argued that the "men
of affairs" thus brought to a debate about the order of nature weighty, albeit contradictory, evidence from the implicit (and explicit) taxonomies
of craft and trade. Looked at this way, the processes by which diverse oils
were manufactured (boiled out, pressed, leached) created meaningful animal "classes" within the circle of oil makers, and in this domain whales
belonged with cattle, and haddock with shark—regardless of the various
manifest dissimilarities within these groups. Where buying and selling

[51] Ibid., p. 53.
[52] Ibid., p. 64.

were concerned, the market generated its own categorical hierarchies, which were continuously affirmed by transactions among interested parties, pored over in daily publications, and enumerated by means of that precise and significant figure called price. Such forms of fine discrimination could not be overlooked in an effort to settle just what whales were, and where they belonged in the order of things.

With considerable sums of money on the line, much could hang on a stray word. Toward the end of the trial Judd's lawyers called James Maurice's former assistant, who gave testimony strongly suggesting that Maurice, the former oil man, had been obliged rapidly to shuffle his own nomenclature after taking up his new commission as the long arm of the fish oil law. Present while Maurice conducted one of his controversial early inspections on several barrels of whale oil, this witness testified that he had overheard someone ask Maurice casually if he was inspecting "fish oil." To which the inspector, "answered no; but afterwards, looking up and correcting himself, said, yes."[53]

[53] Ibid., p. 45. An oil man, Borden Chase, testified to the same effect, p. 46. Note that Anthon alluded to this incident as one where Maurice had been "entrapped, for the moment, by the question artfully put to him" (p. 59).

SIX

The Jury Steps Out

THE KNICKERBOCKERS SLAY A YANKEE WHALE

Reviewing the defendant's case in his closing statement, William Price invoked again the testimony of his client's star witness, Samuel Latham Mitchill, reminding the jury that this man—one of the "luminaries of the present age"—had brought his taxonomic *élan* to the fundamental problem before the court. But, surprisingly, Price then proceeded to draw his listeners' attention not to Mitchill's careful parsing of the creatures of the sea, but rather, first and foremost, to his taxonomy of the creatures of the court and its attendant institutions: "Doctor Mitchill," Price explained, "has aptly *classed the community,* and described the composition of the legislature . . . a representation of the whole community."[1] What remained to the court was the thorny business of weighing the authority and relevance of these different views: the jury would need to decide, above all, *who should decide* whether a whale was a fish.[2]

It has been the aim of this book to expand Mitchill's classification of the relevant "community"—which, as we have seen, he divided into philosophers, whalemen, and men of affairs; and which Sampson expanded with the addition of a fourth catch-all class, "everyone else"—into something like a field guide to the Mayor's Court in late December of 1818. In the process, the preceding chapters have opened the case of *Maurice v. Judd* into an opportunity to examine the place of spouting fish in formal natural history in the period (tracing the fate of footless quadrupeds and pisciform Mammalia through the emergence of the "new philosophy" of comparative anatomy in the late eighteenth and early nineteenth centuries), while also attending to several different significant forms of vernacular expertise concerning these creatures (from the "super-

---

[1] Both quotes: *IWF,* p. 50; emphasis mine.

[2] Hendrik Hartog ("Pigs and Positivism,"*Wisconsin Law Review* 4 [1985], pp. 899–935, section III) examines the question "Who Decides Who Decides?" in the context of a New York Mayor's Court case from 1819; I am here using the same formulation in examining how the process of the trial gave rise to competing accounts not only of the issues at law, but also of the proper means by which those issues ought to be resolved. Legal scholars have generated a large specialist literature on "comparative institutional choice," only some of which engages the historical evolution of U.S. practices.

ficial anatomy" of the blade-wielding whale hunter, to the "craft taxonomy" of the chandler). The circulations of knowledge between and among
these different communities—for instance, whalemen's irreverent commentaries on, and real contributions to, bookish cetology—have made it
clear that the boundaries of Mitchill's human classes were by no means
impermeable, even as such cycles of citation and derision were most often
used to reinforce differences between kinds of knowing and (signally)
kinds of men. Lest "everyone else" be forgotten, an effort has been made
throughout this study to pay particular attention to the most local contexts
of natural history display, pedagogy, and popular publishing in the city of
New York in the first decades of the nineteenth century, and this has
made it possible to sketch a more general picture of the whale as seen
from the ordinary working world of lower Manhattan in these years.

We are now in a position to turn to the resolution of the trial, examining how the lawyers tried to frame the meta-questions that hung over
all the testimony of the case ("Who should decide? How?"), and then,
turning to the jury instructions and the verdict itself, to show how all the
different arguments at play fared on the field of discursive combat. It
will be my argument in this penultimate chapter that, in the end, a re-
gional taxonomy of Americans—a human biogeography of the early
Republic—proved decisive to the resolution of whether whales were fish
in a way that Mitchill's careful taxonomy of the Mammalia, finally, did
not: before the trial ended, a polemical taxonomy of the citizenry had
significantly undermined the authority of a polemical taxonomy of the
cetes. For all his many impassioned efforts throughout his career—in his
Columbia lectures, in his political service, and in his studious efforts to
coin a veritable nomenclature of the nation (Fredes, Fredon, Fredonia)—
to unify the United States with natural history, when he "classed the
community" in *Maurice v. Judd*, Mitchill opened a can of worms that
would bore ugly holes in his carefully arranged book of nature.[3]

WHO DECIDES WHO DECIDES?

Who should decide the case of *Maurice v. Judd*? In one sense the answer
to this question was implicit in the forum: a sworn jury sat in the Mayor's

[3] For a delicious taste of Mitchill's enthusiasm for national nomenclature, see his broadsheet "Generic Names for the Country and People of the United States of America" (no date),
which appears in the "Printed Matter" folder of the Catherine Mitchill Papers in the Library
of Congress. It features exemplary couplets like these: "Their Chiefs, to glory lead on / The
noble sons of Fredon," and "No land so good as Fredon / To scatter grain and seed on." Pity it
never caught on / Much beyond Manhatt-on.

Court and awaited the opportunity to give their verdict on the case. But throughout the trial Judd's lawyers repeatedly suggested that the proper venue to resolve the dispute was not the court at all, but rather the state legislature, which had passed a hopelessly imprecise statute in the first place. As Price put it in his closing arguments: "there is too much doubt and ambiguity to justify a finding upon this evidence . . . so that it will be much the safer course to find here for the defendant, and send the plaintiff back to the legislature, to have its [i.e., the statute's] sense declared or explained."[4] It is important to recognize that this was a standard form of defense argument in this period in the New York courts, and the posture drew on a broader hostile rhetoric toward the intrusions of "nuisance law," which could be presented as particularly dangerous and undemocratic when left in the hands of the judicial branch.[5] Again and again, Judd's attorneys underscored this idea, asserting that the legislation in question—since it hampered the free activity of individuals, and rewarded a state-appointed gadfly (in this case James Maurice) at the expense of truly industrious persons—should be given the narrowest possible compass, until properly democratic institutions explicitly expanded its scope. As Price put it:

> The statute on which this suit is founded, imposes a tax upon commerce, of which freedom is the life, and which never should be subjected to any unnecessary or vexatious restraints. All laws tending to clog or fetter commerce should be construed with extreme strictness, as being against common rights and national policy.[6]

And Bogardus drew out the implication of this position when he posed a rhetorical question to the jurors before resting the case for the defense:

[4] *IWF*, p. 51. See similar arguments by Bogardus: "It is certainly much better, gentlemen of the jury, that your verdict should stay his [Maurice's] hand till the legislature, since the matter is drawn into question, shall have an opportunity to declare its own meaning" (p. 57). A version of this is repeated on p. 58.

[5] For a discussion of this issue, and the related issue of the relationship between the city and state legislative institutions, see Hartog, "Pigs and Positivism," pp. 906–907, and particularly the citations in his notes 29, 32, and 33. Interestingly, Price and Bogardus were here turning against Sampson a rhetoric for which he was well known. He had built part of his argument in the celebrated case of *People v. Melvin* (better known as "The Trial of the Journeymen Cordwainers") on the grounds of this "Jeffersonian distrust" of legal/juridical meddling.

[6] *IWF*, p. 50. This view—that statutes regulating commercial activity ought to be construed narrowly—hardened into formal doctrine by the middle of the century. See: Theodore Sedgwick, *A Treatise on the Rules Which Govern the Interpretation and Application of Statutory and Constitutional Law* (New York: J. S. Voorhies, 1857); and Thomas M. Cooley, *A Treatise on the Constitutional Limitations which Rest upon the Legislative Power of the States of the American Union* (Boston: Little and Brown, 1868).

"why should you, upon uncertain testimony, take money out of the pocket of a respectable commercial man, to put it into the hands of a greedy office holder, who never worked for, or *earned* it by any honourable industry, nor has any meritorious claim to it whatever?"[7]

Sampson and Anthon parried these claims on behalf of their client Maurice in an equally formulaic way, emphasizing that what was at issue was hardly a sallow civic parasite, but rather, as best anyone could tell, a "faithful public servant." Moreover, if punitive fines were to be disallowed because they had not been "earned," then the jury might as well tear up the whole statute book.[8] On the contrary, the plaintiff's lawyers argued, it was the defining characteristic of "well-ordered communities" that legislative energies hunted out every deviation from the "plain path of honest dealing," and that the resulting statutes "must, therefore . . . receive a liberal interpretation to reach the evil."[9] True, Sampson conceded, "[i]f there be any reason why the law should be altered or amended, it concerns the legislature," but that, he reminded the jurors, was an entirely different matter, having nothing to do with the case before them:[10]

> The counsel for the defendant desire you to decline the decision, and thus to send back the statute to the legislature for their interpretation. You cannot safely follow their advice; you have an oath on record which you would thereby violate. You are sworn to decide between these parties, according to the evidence.[11]

The questions at issue—the proper scope of the fish oil statute; the correct construal of the terms "fish oil" and "fish"—belonged, in the final hour, to the jury, as the judge would instruct them at the close of the trial, before sending them away to reach their verdict.

In the endgame, then, *Maurice v. Judd* would hinge on the squeaking and ancient pegs of statutory interpretation. He who succeeded in setting that tricky fulcrum just so would be sure to see the scales of justice fall his way. This was by no means lost on the cagy lawyers, who knew better than anyone that all the evidence they could muster would ultimately hang in the balance at the moment the judge instructed the jury on the proper way to understand the law. For this reason both sides joined early, heatedly, and persistently in a debate over the correct canons

---

[7] *IWF,* p. 55; question mark added. For similar arguments, see also pp. 54 and 57.
[8] Ibid., p. 59.
[9] Ibid., p. 2.
[10] Ibid., p. 76.
[11] Ibid., p. 63.

for the interpretation of the statute, and its proper application to the facts in the case. In fact, at the very opening of the trial, General Bogardus took out a copy of Blackstone's *Commentaries,* and read to the jury a two-page excerpt on the construction of statutes, in order to lay what he took to be the essential foundation for all of the evidence in the case. The result is an invaluable window into what were understood to be the relevant means by which to resolve the controversy.[12] In summary, Blackstone counseled that "The fairest and most rational method to interpret the will of the legislator, is by exploring his intentions at the time when the law was made, by *signs* the most natural and probable."[13]

Responding to the exhortation to ascertain the intent of the legislators, Judd's lawyers went so far as to seek out and call as a witness the actual legislator who had chaired the committee responsible for the fish oil statute: Peter Sharpe, a city representative to Albany. Here was an opportunity to sound legislative intent directly (rather than by questionable "*signs*"), but the court disallowed any part of Sharpe's testimony that purported to provide direct access to the assembly's will in passing the law, thereby hamstringing a potentially decisive defense witness. Sharpe was permitted, however, to dilate on what he "may happen to know as a member of the community," which gave him the opportunity to communicate between the lines of his answers that, as far as he was concerned, the law had been meant to deal with fish oil in the narrowest sense (i.e., liver oil), since that was the kind of oil with which tanners were most concerned, and the oil most commonly coming to purchasers in an unacceptable state.[14] Along the way, Sharpe alluded to his corre-

---

[12] He would return to Blackstone again in his concluding statements (*IWF,* pp. 56–57). On the subject of the status of Blackstone in U.S. jurisprudence in the nineteenth century, see Albert W. Alschuler, "Rediscovering Blackstone," *University of Pennsylvania Law Review* 145, no. 1 (1996), pp. 1–55. Note that Bogardus's actions did attract comment, and it is clear that this much Blackstone was seen to be "curious" by some. See asides in the report on the trial by the *Port Folio* of Philadelphia: "Is a Whale a Fish?" *Port Folio* 8, no. 2, also numbered "no. 224" (1819), pp. 129–133, at p. 129.

[13] *IWF,* p. 22–23 (Bogardus citing Blackstone). As for the nature of these "signs," Blackstone offered the following, as Bogardus explained in a continuation of the given quote: "And these signs are either the words, the context, the subject matter, the effects and consequences, or the spirit and reason of the law." It is interesting to note that this aspect of Blackstone did not fare particularly well in the U.S. over the course of the nineteenth century, since it confronted an emerging republicanism less inclined to understand statutory construction as fundamentally entailed to these kinds of retrospective diagnostic exercises. Sampson himself rejected this aspect of Blackstone altogether in *People v. Melvin.* For an introduction to antebellum ideas about the common law, see David Brion Davis, *Antebellum American Culture: An Interpretive Anthology* (Lexington, MA: Heath, 1979), part 2, chapter 3.

[14] *IWF,* p. 24 ("The Recorder [i.e., the judge].—What the witness may happen to know as a member of the community, he may state, but not what he knows as a member of the legisla-

spondence with Gideon Lee, "with whom," he noted, "the law origi-nated," and to the other letters received at the "time of the origin and pendency of that bill," in order to make it clear (tacitly) that, having watched the emergence of the statute at close range, he thought it had no bearing on spermaceti oil.[15]

Again and again Judd's attorneys returned to this issue of legislative intent, pressing home the idea that the law had been tailored to address a specific problem, a problem unknown among purveyors and consumers of spermaceti and whale oil. As Bogardus put it:

> You will easily see that this law was only intended to prevent
> some fraud that *existed,* and that where none existed, it could
> have no application. It can have none in this case for there were
> no frauds in the article of sperm oil, nor indeed in any whale oil.
> The evidence, or the weight of the evidence, goes to show that
> these frauds were only in the fish or liver oil, and as *they* only
> called for inspection, to them only can the law be applied.[16]

If it had been otherwise, the defense went on to contend, why had the plaintiff's lawyers objected so vigorously to the testimony of Peter Sharpe, who had come to the court to dispel any doubt concerning the import and intended scope of the legislation?[17]

Against these arguments Sampson and Anthon set the testimony of Gideon Lee himself, who—though also prohibited from speaking

---

ture"), It should be emphasized that the whole issue of the proper relationship between the law-making power of the state legislature and the traditions of local municipal authority was very much in flux in the early nineteenth century in New York City. In the context of these on-going debates, a law like the "act for guaging [*sic*] and inspecting of fish oil" (and Lee's lobby-ing efforts) might perhaps be understood as part of a relatively new (but increasingly impor-tant) strategy for bypassing the Corporation of the City of New York (as well as the customs and traditions of the merchant communities there), in order to make and enforce regulations from Albany. For a discussion of these dynamics in the period, see Hendrik Hartog, *Public Property and Private Power: The Corporation of the City of New York in American Law, 1730–1870* (Chapel Hill: University of North Carolina Press, 1983), particularly chapters 9 and 10.

[15] *IWF,* p. 25. Teasing out one of Blackstone's "signs," Price noted to the court that while the *title* of the statute mentioned "fish oils" in the plural, the *body* of the statute referred to "fish oil" in the singular; he intimated that this detail spoke to his preferred (limited) construal of the law. Anthon later picked up on this question, p. 63.

[16] *IWF,* p. 56.

[17] Ibid., p. 50. See Price: "Mr. Sharpe's explanation of the motives and intentions of the legislature in passing this law, and of the objects it was meant to embrace, were opposed by the plaintiff's counsel, although he was the person who could best have informed you upon that subject, having been the chairman of the committee to whom it was referred, and who reported on the bill. [¶] This may serve to show you, gentlemen, the fears and wishes of the counsel, and how anxious they were to suppress the evidence that would have best enlightened your judgements."

directly to the matter of the intent of the framers of the act (though no legislator, he was, after all, the bill's recognized author)—had also managed to lard his testimony with unmistakable references to his own intentions in drafting the legislation, which were very much at odds with the position taken by Sharpe.[18] Summing up, Sampson addressed the matter directly:

> As to the evidence touching the meaning of the legislature, I had great doubts whether it was admissible; but as the counsel for the defendant introduced it, we have rebutted it, I think successfully. We have produced Mr. Lee, a competent and most intelligent witness; he drew the bill, and tells you with what view the act was solicited from the legislature, what mischiefs it was intended to remedy; and that the term, *"fish oils,"* was used as the most certain to embrace all marine oils, and that of whales, amongst the rest, without any possibility of doubt or equivocation.[19]

In the end, because these direct efforts to glean the legislative intent behind the act in some sense cancelled each other out, the case would ultimately hang on clues as to the proper construal of the terms "fish" and "fish oil(s)." Much evidence had of course been heard on this topic. How were the jurors to choose among the different accounts offered by practical whalemen, philosophers of the natural order, and respectable men of affairs? On this decisive problem, Bogardus offered the following principles, again quoted from the *Commentaries* of Blackstone: "Words are generally to be understood in their usual and most known signification; not so much regarding the propriety of grammar, as their general and popular use." This held for all ordinary terms, but "terms of art, or technical terms, must be taken according to the acceptation of the learned in each art, trade, and science."[20] "Reason" was to be the infallible last resort of jurors and jurists alike when they faced language that defied these principles, or that set them at cross-purposes.

Was the term "fish," then, "a term of art," calling for the gloss of men of science and technical expertise? Or was it, rather, a common term, to be understood in its popular sense, as used by members of the community? In instructing the jury, the judge, Richard Riker, made it clear that they could opt for the former if they chose: "If then, you are of opinion

[18]Lee was asked directly to state his intent in using the term "fish oils" when he drafted the statute, but the judge sustained an objection by the defendant's counsel (*IWF,* p. 42).

[19]*IWF,* p. 64.

[20]Ibid., p. 23.

with the naturalists, that a whale is not a fish . . . you may then find for the defendant, as he will not have incurred the penalty given by this statute."[21] Predictably, William Price, closing his defense of Samuel Judd, left no doubt that this was the proper path: "since all legislative acts must be construed according to *sound reason*," and since science was the elaboration of reason as it applied to nature, it must be to science that the jurors applied for the "most convincing and authoritative evidence of what the legislature must have intended."[22] And thus the case was an easy one, since

> to whom could you apply with more safety, on a subject con-
> nected with science, than Dr. Mitchill; a man abounding in all
> useful knowledge . . . who, to use his own language, has stood
> upon the shoulders of those that went before him; who gathers
> his information from the scriptures no less than from all other
> ancient and modern authority, and out of the abundance of this
> knowledge, and the fulness of his conviction, solemnly swears
> that a whale is not a fish.[23]

Here was "the first man of his age, and country," who had spelled out the solution to the whole case; Sampson and Anthon had been forced to truck out dubious and antiquated sources in their effort to undermine Mitchill's assurances, for otherwise "why did they not produce other learned men to oppose his opinions?"[24] Not that it really mattered: the jury could safely put aside the library of learned authorities that had been cited on both sides, since Mitchill, "in himself, is greater than them all."[25] If, therefore, reason was to be the balance on which the question was weighed, Mitchill—who in his person embodied nothing less than reason itself—could be trusted to try the issue in his bare hands, alone.

But was this a matter to be entrusted to the analytic balance of natural history? Anthon, closing the case for the plaintiff, returned to the canons of statutory interpretation to argue that the meaning of the term "fish" belonged not to the schoolmen, but to "the whole community":

> As to the opinions and classification of modern naturalists on
> the subject, they are undoubtedly against us as far as their evi-
> dence has been heard. And if this statute is to be interpreted in

[21] Ibid., p. 78. NB: I give and gloss Riker's full instructions below, at footnote 39.
[22] Ibid., p. 51; emphasis mine.
[23] Ibid.
[24] Ibid., p. 52; question mark added.
[25] Ibid.

this way, we should be bound to surrender our common sense, and say with Doctor Mitchill, that a whale is not a fish, and that whale oil is, consequently, not fish oil. But, gentlemen, such a construction of a statute would be in opposition to the well established rules of the common law. *Statutes being enacted to regulate the conduct of the whole community, the words of the statute are to be interpreted according to their common usage and acceptation.*[26]

What, then, constituted "the whole community"? And what should the jury make of the many non-scientific witnesses—respectable men like Fish, Hazard, Comstock, Chase, and others—all of whom swore that whale oil was not fish oil, and most of whom also insisted that whales were not, properly speaking, fish? Here Sampson and Anthon unveiled a final strategy whose groundwork they had laid throughout the evidentiary phase of the trial—a strategy that positioned them to argue that these opponents were not, in fact, members of the relevant community.

As Anthon acknowledged to the jury, they had indeed seen a set of witnesses testify that "fish oil embraces exclusively the cod liver oil" and insist that this was the "common commercial acceptation of the words." However,

It very speedily became manifest, that all these gentlemen were of *eastern origin,* and were giving to us the acceptation of the words in the *eastern states only.* When this became apparent, the question was reiterated on our part to every witness, who maintained that opinion, whether he was not from the eastward, and the invariable affirmative demonstrated the force of our inquiry.[27]

So *here* was the true explanation for the clique of whale-not-a-fish outliers. Yes, these were, by and large, oil men, with their narrowly self-serving preoccupation with avoiding the inspection fees. But perhaps this did not go to the heart of the matter (after all, the tanners, admittedly, had their private interests at stake as well). Yet there loomed a still larger and more urgent rationale beneath the improbable taxonomy of Judd's fifth-column witnesses. A secret confraternity had been giving silent shape to the case: these anomalous oath-takers, who had outrageously and vociferously vouched for mammalian cetes, were *not really New Yorkers.* They were, rather, "Easterners"—Boston men, long-boned New Englanders, specimens of the sanctimonious, tight-fisted, and well-to-do *arrivistes*

---

[26] Ibid., p. 60; emphasis mine.
[27] Ibid., p. 62; emphasis mine.

seen more and more in the mid-Atlantic marches to the south of their natural habitat. Once this had been revealed, it became clear that those who wished the whale no fish had in fact been attempting, covertly, to import an *alien usage,* one that could have no place in the proper construal of a New York statute. As Sampson put it, it would hardly be appropriate to permit "the particular phraseology of a small section of the union" to "control the general acceptation of words."[28] And Anthon dismissed the notion of non-fish-whale oil as nothing more than a "mere provincial usage" of New England, one that could readily be ignored.[29]

It was true that during the course of the trial Anthon and Sampson had carefully elicited information from witnesses about their origins, and the number of years they had resided in New York. The inspiration for this maneuver may have come from an aside by the Swamp prince Israel Corse, who hailed from Maryland himself, and who on the stand commented in passing that he "believed that the eastern people made some distinctions about fish oils" but such curious locutions were not known "to the southward, or anywhere but to the eastward."[30] No sooner had the point been observed than one by one all of Judd's witnesses were queried about their trajectories to New York: Captain Fish, from New Bedford, acknowledged that he was "an eastern man" (though he protested that he had been for twenty years in New York); the oil merchant Thomas Hazard, it turned out, "had followed the business of whale oil twenty-seven years to the eastward, and three here"; the chandler Borden Chase "had lived seven years to the eastward"; and Nathan Comstock was obliged to admit that he "formerly lived at Nantucket, where they are more acquainted with whale than fish oil."[31] The former gauger, Ralph Duncan, confirmed the emerging picture, explaining that, as far as the distinction in the name of liver and whale oil was concerned, "the merchants who brought the oil made some such distinction, but they were mostly eastern men."[32]

It was a subtle ploy, but an effective one, playing brilliantly as it did on the Yankee-Knickerbocker rivalry that had stewed New York elites

[28] Ibid., p. 64.

[29] Ibid., p. 63.

[30] Ibid., p. 44.

[31] Ibid., pp. 45–47.

[32] Ibid., p. 45. Of course, Gideon Lee—who was, after all, originally from Worthington, Massachusetts—could easily have been assigned the same geographical designation, but he steadfastly rejected the notion, introduced in cross-examination, that he too was an "eastern man," asserting instead that he was a "northern man," and that Worthington was "53 miles from Albany" (p. 43).

for a generation.[33] From the perspective of the early twenty-first century, the modern reader is likely to think of the "Yankees" as emphatically native to the Big Apple—every bit as native as the "Knicks." But those pinstriped uniforms in the Bronx camouflage what was originally an invasive species: in the early nineteenth century "Yankee" was an epithet in New York City, meaning as it did a *New Englander*—someone without the Dutch blood that distinguished the true citizen of the city once known as "New Amsterdam."

Judd's lawyers, acutely aware that a well-placed lever had been pulled, rose instantly to denounce the low tactics of the opposition, and petitioned the jury to reject a specious and ungracious tack in the proceedings. "I trust," Price intoned hopefully, that the opposing counsel "will gain nothing by illiberal allusions, and by fixing on respectable and honorable citizens, the contemptuous epithet of yankees." Such "unjust appellations," hurled at some of "the most enterprising, spirited, and useful of our citizens," represented bare-knuckle lawyering at its worst.[34] To which Anthon, in wide-eyed innocence, replied that his side had done no name-calling at all. Indeed, he and his colleagues could not agree more about the virtues of hard-working Yankees: "We respect this class of our fellow citizens as much as they [the defendants] can," Anthon explained peaceably, before twisting the blade: "but we are not, therefore, willing that they should dictate to us the meaning of words."[35]

Price and Bogardus did what they could to address the geography and sociology implicit in this clever contention, noting, for instance, that if the law had really been meant to cover whale oils, then it was impossible to understand why no inspector had been assigned to Sag Harbor, the state's only great "eastern" whaling depot (the legislation called for the commissioning of three inspectors for the whole state, one for Al-

---

[33] For a concise summary of the conflict in New York City, see Burrows and Wallace, *Gotham*, pp. 452–456. The classic treatment of the eighteenth-century roots of the issue (in land tenure disputes and agrarian reforms in the state as a whole) remains Dixon Ryan Fox, *Yankees and Yorkers* (New York: New York University Press, 1940). For the cultural dimensions, see also Bender, *New York Intellect*, pp. 131–133. Bender shows that part of the conflict was about perceived intellectual habits: Boston "thought" proper was to be contrasted with Knickerbocker "wit, satire, and burlesque." The term "Knickerbocker" itself, of course, only dated to Washington Irving's pseudonymous publication of *A History of New York* (New York: Inskeep and Bradford, 1809) under the name "Diedrich Knickerbocker," but the actual rivalry was considerably older. For Knickerbocker literary culture, see Perry Miller, *The Raven and the Whale: Poe, Melville, and the New York Literary Scene* (Baltimore: Johns Hopkins University Press, 1997 [1956]), particularly book 1.

[34] *IWF*, p. 53.

[35] Ibid., p. 62.

bany, one for Troy, and one for "the city of New-York, whose powers shall extend to, and include the village of Brooklyn").[36] In addition, Price and Bogardus, their backs against the wall, undertook their own effort to delimit the relevant community, though by means of class distinctions rather than sectional antipathies: they reminded jurors that in weighing the testimony from the practical whalemen, they would do well to recall that *Captain* (whale-not-a-fish) Fish had been "master of a vessel in the trade," whereas *Mr.* (whale-a-fish-far-as-I-know) Reeves "was but a common whaler who had made but three voyages" before the mast. "In which," Price asked, "can you best place your confidence?"[37] But there was an *ad hoc* quality about these points, a sense that the defense was grabbing at straws.

The Yankee whale, by contrast, had become a fat and easy target in a Knickerbocker court.

Framing the case for the jurors at the end of the closing statements, the honorable Richard Riker made it clear that the case came down to the proper interpretation of the statutory phrase "fish oil," and he provided two ways they might attack this problem. On the one hand, there were those creatures out there in the sea: "The oil in question was called spermaceti," he pointed out, because it "is extracted from an animal called the spermaceti whale." Thus, he continued, "there comes the question, is that whale a fish?" Riker then rehearsed for the jurors the reasoning on each side of this problem, and conveyed it to their hands without tipping his cards: "There is much plausibility in the arguments applied to the subject in these various points of view; and you must decide amongst these conflicting opinions." On the other hand, however, Riker also offered the jury a parallel path to the resolution of the case: as they had heard, the term "fish oil" might be approached as a particular term in commerce, where it appeared to mean either all marine oils, or, perhaps, exclusively oil from the livers of certain fish; the jury might thus judge which of these meanings was correct and relevant, using the evidence they had heard on the subjects of kegs and consumers, and putting aside the bodily business of whales and fish themselves.[38]

Finally, having defined this fundamental fork in how the case could be decided—either via the question of whether whales were fish, or via

[36] This argument appears in *IWF*, p. 58 (see also p. 21).

[37] *IWF*, pp. 53 and 58; question mark added in this quote.

[38] Riker's jury instructions in *IWF*, pp. 76–78; quotes are from pp. 76 and 77.

the question of whether whale oil was fish oil—Riker left the choice of route to the jury, explaining to them that they were free to choose which way they reasoned about the problem. At the same time, he crafted a parting instruction very favorable to Samuel Judd, explaining that *either* a finding that whales were not fish *or* a finding that whale oil was not fish oil *necessitated* an acquittal, and that, moreover, if they entertained doubts, they were free to dismiss the whole affair.[39] With that, the jury retired.

Fifteen minutes later they returned, and announced a verdict for the plaintiff: Judd, Mitchill, and the Yankee whale had been routed.

Picking Up the Pisces

Word of the verdict made the rounds in the daily papers immediately, and New Yorkers awoke to the new year to learn that, as the *New-York Gazette* announced prominently,

> The great trial between Mr. Maurice inspector of oil, and a gentleman who bought three barrels of *whale* oil without in-spection was brought to a close last evening, after occupying the court for three days. The Jury found a verdict for Mr. Maurice, having decided that a *whale* is a *fish* and whale oil fish oil.[40]

And the *Evening Post* glossed the happening with satirical glee. The "learned Dr. Mitchill" had failed in his quixotic enterprise, the paper re-ported, since in the face of his elaborate arguments, "the jury, by their

---

[39] To be precise about the final portion of the jury instructions, Riker's words were recorded as follows: "If, then, you are of opinion with the naturalists, that a whale is not a fish, or if you think that, in the commercial sense, fish oil does not mean whale oil, you may then find for the defendant, as he will not have incurred the penalty given by this statute" (*IWF*, p. 78). The "may" here is properly read as a "must" or a "shall" (that is, as an obligation) on the following grounds: since Riker states explicitly that if the jury finds A (whale not a fish in sense of naturalists) or B (whale oil not fish oil in commercial sense) then "he [the defendant, Judd] *will not have incurred the penalty given by this statute*," any reading of "may" here short of an implication of obligation would have Riker instructing the jury that they possessed the discretion to punish a man who had not broken the law. But this is impossible, even by the somewhat loose standards of courtroom practice of the day. For those not entirely satisfied by this (admittedly lawyerly) construal of the judge's language (or for those who would insist, not unfairly, that juries seldom perform such mincing analysis), it is worth noting that the jury's verdict was widely interpreted (in the press, etc.) as a rejection *both* of Mitchill's non-fish whale *and* of whale oil as something other than a kind of fish oil (see, for instance, *The New-York Gazette,* 1 January 1819 [misprinted as 1818], p. 2, quoted at the opening of the next section). As far as Riker's concluding comments concerning doubt, they were as follows: "You may also, if you doubt, this being a penal act, and against the common right of the citizen, which is to deal in all commodities not noxious, or restrained or prohibited by law, find against the plaintiff." (Here too "may" seems to carry the force of obligation, since juror doubt in a criminal case ought to bar conviction.)

[40] *The New-York Gazette* 1 January 1819 [misprinted as 1818], p. 2; emphasis original.

verdict, decided that *a whale is a fish*," a finding that another paper likened to an elaborate and superfluous legal affirmation of the garden-variety tautology that "rhubarb is rhubarb." A few days later the *Christian Messenger* of Vermont put its finger directly on the science-versus-the-people aspect of the case, summing up the positions represented in the trial thus: "Counsellor Sampson supported the *popular*, and Dr. Mitchill the *philosophical* side of the question."[41] The verdict, by these lights, struck a blow for plain folk everywhere. Soon periodicals and newspapers from Maine to Louisiana, Kentucky to South Carolina had picked up the story, and sketched the contours of the trial with irreverent delight, several of them likening Mitchill's discomfiture to that of Joseph Banks as depicted in the notorious satire of John Wolcot (alias Peter Pindar) entitled *The Lousiad* (1785). This ribald send-up of court culture around George III—departing from the report of a louse found on the King's plate—contains a memorable scene set among the savants of the Royal Society, who, under Bank's lead, endeavor to demonstrate that the louse is in fact a species of terrestrial lobster. Their experimentation involves boiling countless lice in an effort to establish that they too turned red when cooked.[42] *Niles' Register*, and several other papers, reporting on the undoing of the "sapient Dr. Mitchill," recalled Pindar's doggerel as the final word on Mitchill's delusional taxonomy: it too had foundered on the fixed reef of fact, a fact as plain as the nose on the doctor's very familiar face. As the *Register* put it:

> Sir Joseph Banks, we are told by the facetious Peter Pindar, once made an experiment, to satisfy himself whether fleas were not lobsters, by boiling them to see if they would turn red, but the result disappointing his expectations, he is made by the poet to exclaim, peevishly,

> "Fleas are not lobsters, d——n their souls."[43]

Mitchill, it was to be supposed, found himself similarly cursing the cetes in the wake of *Maurice v. Judd*. But perhaps the best line of all was reserved to the *Evening Post*, which found occasion to muse thoughtfully, "Pray sir, how goes it with whale oil now? Is it oil of fish, or of flesh, or of *red herring*?"[44]

[41]"Rhubarb is rhubarb" comes from the *National Advocate*, 1 January 1819, p. 2. *Christian Messenger*, 13 January 1819, p. 3; emphasis added.

[42]The original ode (canto 3) reads in relevant part, "'I've boiled just fifteen hundred,' Jonas whined, / 'The devil a one changed colour I could find' / 'How,' roared the President and backwards fell, / 'There goes then my hypothesis to hell!'"

[43]*Niles' Weekly Register*, 21 August 1819, p. 31.

[44]*Evening Post*, 10 August 1820, p. 2; emphasis added.

For the *Lady's and Gentleman's Weekly Literary Museum and Musical Magazine* of Philadelphia there was even a pleasing symmetry between the trial of the whale by means of biblical citations to Jonah and the trial of Jonah by means of the biblical whale. After all, the paper noted, the trial itself had lasted exactly three days, the precise period that Jonah found himself in the belly of the beast; in New York City the beast had been swallowed by the courts, only to be vomited up three days later, chastened, but none the worse for wear, since "Finally, the whale was permitted by the jury to continue king of the scaly tribe, in despite of the theological and ingenuous arguments of Dr. Mitchill."[45]

Even as these flippant voices rose to hail the legal affirmation of common sense and common usage, a second chorus, or perhaps a mumble, sounded a lower note. As men like Henry Meigs circulated word of the court's decision among the learned circles of Philadelphia and Boston, heads collectively shook at this new performance of New York's stridently philistine culture. Whales as fish? The better sort knew better. It was in rejoinder to such sotto voce complaints about the outcome of the case that Sampson claimed to have decided to publish his account of the trial, since, as he put it, in the aftermath of *Maurice v. Judd* he had been forced to watch as "philosophers of the new school have flown to arms, and already brandish their pens in deadly defiance of the Knicker-bockers."[46] If they wished to lament the state of science in Manhattan, they would do well to study what had actually transpired in the case, he explained, rather than seizing on "loose rumors and running reports, and . . . tattling gossip."[47] By August, when Sampson's fifty-cent pamphlet had appeared in book shops, a second round of commentary on the trial had begun, occasioned in part by a lengthy excerpt of Mitchill's testimony which appeared in Philadelphia's notoriously Federalist journal of belles-lettres and political commentary, the *Port Folio*.[48] The anonymous reviewer there viewed the whole incident as exemplary of the circus-like world of Philadelphia's sister city to the north:

[45] *Lady's and Gentleman's Weekly Literary Museum and Musical Magazine* 3, no. 18 (22 Febuary 1819), p. 144. The editors were permitting themselves some poetic license here, since the trial actually only lasted for two days.

[46] *IWF*, p. vi.

[47] Ibid., p. v.

[48] It has been argued that Joseph Dennie's *Port Folio* represented the fruit of a Federalist retreat from the stormy waters of American politics into the narrow bay of literary commentary. See William C. Dowling, *Literary Federalism in the Age of Jefferson* (Columbia: University of South Carolina Press, 1999). Note that Dennie died in 1812, after which the journal was edited by Paul Allen, Nicholas Biddle, Dr. Charles Caldwell, Thomas Cooper, Judge Workman, John Elihu Hall, and his brothers. See Albert Henry Smyth, *Philadelphia Magazines and Their Contributors, 1741–1850* (Philadelphia: R. M. Lindsay, 1892).

There are many strange fish in New-York, as some of our readers
may know, who never heard of this singular trial. They are not,
however, inhabitants of the watery element, but may be found on
the Battery, in Broadway, or at the sittings of the *Institute*.[49]

The allusion to the New-York Institution (i.e., the "Institute") merely set
up its high priest, Mitchill, for his fall, since, as the *Port Folio* duly
reported, "the jury, after full deliberation on the theories of the naturalists, the popular acceptation of the terms used in the law, and the rules
of the common and statute law, found a verdict for the plaintiff; thereby
establishing the fact that a whale is a fish, Dr. Mitchill to the contrary
notwithstanding."[50]

From other quarters came still more spirited mockery. New York's
own satirical duo, Joseph Rodman Drake and Fitz-Greene Halleck—
who under the collective pseudonym of "The Croaker" together published rapier-sharp poems on the issues and figures of the day—trained
their blades on Mitchill in March of 1819, publishing in the *Evening Post*
a scalding paean to the Sage of Fredonia, which did not neglect to take
a well-placed dig at *Maurice v. Judd*:

> Oh Mitchell [*sic*], lord of granite flints,
> DOCTUS, in law—and wholesome dishes;
> Protector of the patent splints,
> The foe of whales—the friend of fishes . . .[51]

It is known that a copy of this poem reached the doctor still wet from
the press, and—as John W. Francis remembered the incident—Mitchill,
reading it, looked as if "an arrow [had] just pierced his intercostals." Nor
was his discomfiture short-lived, Francis recalled: "his feelings suffered
annoyance for a long period."[52] The anecdote offers some insight into
Mitchill's own response to the very public rejection of his efforts in
Maurice v. Judd, as does a glance at the legacy of his preoccupation with
the question at issue in the trial. Documentary evidence reveals that
Mitchill would not let the question of the taxonomy of the whale drop
over the next decade, nor did the city let him put it to rest.

[49]"Is a Whale a Fish?" p. 129.

[50]Ibid., p. 138.

[51]"To the Surgeon General of the State of New York," in Drake, *Poems*, p. 8. See also the
edited reprint edition of the collected Croaker poems, which includes some helpful annotations: Joseph Rodman Drake, *The Croakers* (New York: The Bradford Club, 1860). Note that
Drake made fun of Mitchill's whale taxonomy yet again in "To Prince Croaker," also composed
in early 1819.

[52]Francis, *Reminiscences of Samuel Latham Mitchill*, pp. 18–19.

Public interest in cetaceans reached a kind of fever pitch in the wake of the trial, so much so that an anonymous clique of flimflam artists conspired to fabricate (out of what? it is difficult to imagine. . .) a whale-like thing which they promptly took on the road, showing it (by admission only, and doubtless in a poorly lighted tent) in New York City, Albany, and other points upstate. The scam went smoothly until "some discerning persons at Waterloo" got wise and "discovered the cheat." What followed may be the only burning of a whale in effigy in modern history. As one paper explained: "the *counterfeit whale* was set on fire, and the rogues decamped."[53]

But why mess about with smoke and mirrors when the real thing was right offshore? In May of 1820 fishermen working the productive banks of Sandy Hook, New Jersey, spotted a whale spouting in the shallows and, exerting themselves with the tools at hand, managed to subdue the creature and bring it to shore, apparently to Brooklyn.[54] Given the events of *Maurice v. Judd,* the value of the beast was seen to lie more in its mysterious anatomical integrity than in its comparatively unmysterious commercial derivatives, and before the carcass could be wholly disarticulated it was acquired by an enterprising showman and tavern-keeper, G. C. Langdon, who immediately invited Mitchill out to give it his perusal. As cries from neighbors about the stench (and the unhygienic presence of dozens of tons of rotting meat) besieged Langdon's impromptu program for pedagogy and profit, Mitchill himself took pen in hand to draw public attention to this most fortuitous, if tardy, witness for the defense. In the *New-York Literary Journal and Belles-Lettres Repository* Mitchill offered a lengthy rebuttal to his chattering critics, addressed as an open letter to Langdon:

> Sir—There is an excellent opportunity now afforded by your successful exertions, to settle the question lately agitated in New-York and Albany, whether a whale is a fish. The creature killed on Saturday last . . . is of the cetaceous order, and of the kind called *Balano,* by naturalists, or the *bone* whale, or the *right* whale, by people in general.

[53] *Carolina Centinel,* 20 March 1819, p. 4, emphasis original.

[54] There had been, in the eighteenth century, a developed shore whaling community along the New Jersey coast, in addition to a number of whaling ships that sailed from Newark. On the former, see primary source references to the "Whale-men" of Cape May and Sandy Hook in William A. Whitehead et al., eds., *Documents Relating to the Colonial History of the State of New Jersey* (Patterson, NJ: The Press, 1894) vol. II, p. 46, and vol. 19, p. 332 and passim. On the latter: Weiss, *Whaling in New Jersey.* My sense is that in the second decade of the nineteenth century whales in this area were only taken adventitiously by fishermen otherwise employed.

Mitchill then proceeded to enumerate the features of the animal that should be drawn to the attention of everyone who attended "the exhibition," and who wished to know once and for all "IS A WHALE A FISH?" Rightly understood, these features would be, he promised, "sufficient to put the question forever at rest." To that end Mitchill exhorted the thousands of curious visitors to the whale to examine: first, the *ear* (which, though not external, still featured a "meatus, or passage to the internal organ of hearing," a guarantee that this creature "resembles other mammiferous beings"); second, the *eye* (which displayed "eye-lids like those of land animals" as "distinct as those of the cow or the horse" and wholly unlike the "naked eyes" of a fish); and third, the famous arm-like *fins,* which were veritably—as he had asserted on the stand—a man's hand in mittens.[55] As Mitchill explained:

> Fish have fins with rays of bones, giving them a *fan-like* appearance. But the whale has no fins with radiating bones running through them. They are, on the contrary, in the nature of arms in men, or of fore legs in beasts. There is a near resemblance in the organization of the two pectoral appendages (fins as they are called) to the arm of a man; there being a shoulder blade, humerus cubit, wrist and imperfect hand, all invested in one common covering of skin, as a man's hand is if enwrapped in a mitten or close glove.

Mitchill closed by giving Langdon the surgeon-general's vote of confidence on the question of the public welfare ("when I was at the place where it lies, I found nothing to alarm me on the score either of comfort or health," though the "free application of lime" and continued cool weather would be required to keep the show going), and he further praised Langdon's initiative in "contributing to the study of zoology" and promoting "researches in science," while granting "the wishes of the citizens to gratify their rational curiosity."[56] By the end of the month the carcass—which now stank in good earnest—had attracted a deputation of dignitaries, including the Mayor of New York (Cadwallader D. Colden) and General Jacob Brown, the hero of the War of 1812 (who made the outing with his wife). And the newspapers had a field day with

[55] I base the assertion that the whale attracted "thousands" of visitors on a comment (*Portsmouth Oracle,* 27 May 1820, p. 3) that Langdon raised $1,300 in a *single day* from the whale exhibit. Even if this is a gross exaggeration, it suggests a huge audience, particularly since I doubt he could have charged more than twenty-five cents for adult admission.

[56] All quotes from *New-York Literary Journal and Belles-Lettres Repository* 3, no. 1 (May, 1820), pp. 60–61.

the occasion, wondering whether it might make sense to retry *Maurice v. Judd* with Langdon, the Brooklyn publican, as "umpire." Oh, and, if the whale had ears and eyelids (as the doctor promised), did it also have, perhaps, *horns?* It was clearly up to the "lovers of Zoology to visit the Whale and decide for themselves."[57]

As it happened, those who missed Langdon's whale would get other chances: July of 1820 found Mitchill performing a full anatomical examination of a large "Delphinus rostratus, or Herring hog" in the city; and two years later, in March of 1822, Mitchill undertook a public dissection in Manhattan on a small rorqual (the modern term for whales with striated pleats on the lower jaw, like minke, fin, and blue whales) secured in local waters, and exhibited by strong-stomached purveyors of public entertainment.[58] It is in the context of these highly visible anatomy lessons that Drake's snide allusion to Mitchill in a later poem, *The Great Moral Picture,* must be understood. For there Drake puckishly recalled the good old days, "when Doctor Mitchill's word was law," and when "monsters and whales" needed only "a letter from his ready hand / to be the theme and wonder of the land."[59] So regular had such impromptu anatomical carnivals become that when, in 1825, a sixty-one-foot whale was captured in the waters to the south of the city, the *New-York Mirror, and Ladies' Literary Gazette* reported the catch with a nod and a wink in the doctor's direction, noting: "It is supposed that this 'animal,' or 'bird,' or 'fish,' (for it is uncertain in Dr. Mitchill's mind to what species it belongs,) will produce at least forty barrels of oil."[60] And the next year, when a tall tale from the whaling industry made the rounds in the papers (involving a whaleman who had gotten tangled in the harpoon line and ended up rid-

---

[57] On the visit by various worthies: *Evening Post,* 30 May 1820, p. 3. For the ribbing of Mitchill and Langdon both: *National Advocate,* 23 May 1820, p. 2.

[58] This small whale appears to have been a minke whale, since it was identified (by DeKay) as a "*Balenopterus acuto rostratus.*" Both incidents are mentioned in the manuscript Minutes of the Lyceum of Natural History, collection of the New York Academy of Sciences: 25 March 1822 (for the whale), and 24 July 1820 (for the dolphin). It is unclear how public the dolphin dissection was, but there is every reason to suspect that it was done in or near the fish market. There are additional discussions of the 1822 whale display and dissection in *The American Journal of Science and Arts* 6, no. 2 (1823), pp. 364 and 365. The term "rorqual" was used in English in the early nineteenth century, but apparently it was used as a specific name for what Linnaeus called the *Balaena musculus.* See the interesting pamphlet: Patrick Neill, *Some Account of a Fin-Whale, Stranded near Alloa* (Edinburgh[?]: n.p., circa 1810). See also Frederick John Knox, *Account of the Rorqual, the Skeleton of Which is Now Exhibiting in the Great Rooms of the Royal Institution* (Edinburgh: A. Balfour and Co., 1835).

[59] See Fairchild, *History of the New York Academy of Sciences,* p. 61.

[60] *The New-York Mirror, and Ladies' Literary Gazette* 3, no. 10 (1 October 1825), p. 79.

ing his mount like a bronco), the yarn was reported with an elbow in the ribs: "The following would be a fine fish-story, were it not that the famous Dr. Mitchell [*sic*] has proved beyond all controversy, that 'a whale is not a fish.'"[61]

And on it went. A few years later, in 1829, the Boston schoolmaster and city father Ebenezer Bailey, confronting one of the stuffed "mermaids" that enjoyed a minor vogue among hucksters and curio dealers at mid-century (they were the taxidermic offspring of a monkey and a fish, made in Japan and brought back by American sailors), could not resist invoking *Maurice v. Judd* in the closing line of his satiric ode "Address to the Mermaid":

> But go in peace, thou thing of "shreds and patches"—
> Go not, howe'er where Doctor Mitchill is;
> For he will mangle thee, if he but catches
> A glimpse of thy uncouth and monkey phiz,
> And then will swear, in spite of thy long tail,
> Thou art no more a fish than was his whale![62]

In these ways and others *Maurice v. Judd* percolated in the collective consciousness of the young Republic throughout Mitchill's lifetime, and beyond: in 1859 John W. Francis could recall the "still well-remembered case of 'a whale is not a fish'" (giving it a distinctly Mitchill-esque title), and as late as 1907 the court stenographers of the city of New York could remember as a national *"cause célèbre"* the notorious trial—"famous throughout a smiling country as 'Is a Whale a Fish?'"— that set "religion and tradition *versus* iconoclastic science."[63] Nor was the attention limited to the United States. Somehow—perhaps via the circulation of Sampson's pamphlet, or the London publication of the *Port Folio* (or maybe even via Mitchill's direct correspondence with Cuvier or Lacépède)—accounts of the trial crossed the Atlantic, and occasioned critical commentary by practicing zoologists who viewed New York from afar. A long note in

----

[61] *Vermont Gazette*, 4 April 1826, p. 3

[62] See Samuel Kettell, *Specimens of American Poetry* (Boston: S. G. Goodrich, 1829). There is a minor literature on the mermaid hoax. A point of departure is Phineas T. Barnum, *The Life of P. T. Barnum* (New York: Redfield, 1855). See also: Neil Harris, *Humbug: The Art of P. T. Barnum* (New York: Little, Brown, 1973); and Jean Ashton, "Tall Tales and Whales: Wonders of Barnum's Museum," *Imprint* 16, no. 2 (Autumn 1991), pp. 15–25. "Shreds and patches" is from *Hamlet*, act 3, scene 4, and refers to the king's ghost. "Phiz" (from "physiognomy") was period slang for "face" or, perhaps better, "mug."

[63] See: Francis, *Reminiscences of Samuel Latham Mitchill*, p. 23; and Charles Beale, *William Sampson, Lawyer and Stenographer*, p. 24.

Edward Griffith's English edition of Cuvier's *Règne Animal* cited the case as evidence for the continued "confusion of ideas produced" by "misapplying the name of fish to the cetaceous animals":

> A few years ago there was a trial at New York upon this subject: the payment of duty upon whale oil being resisted on the ground that the words of the law were confined to oils produced from fish, and as whale was not fish, the oil from that animal, it was contended, did not come within the letter of the law. To support the allegation, naturalists of ability and Professors of the University gave it in evidence that whale were not fish; but the jury would not be convinced by the learned distinctions of science.

The story occasioned a comment on the broader social, political, and legal importance of the precise classifications found in the science of natural history, and a call for a convergence between this formal language of natural order and the formal language of social order:

> It might, however, be as well a question of investigation with legislators, when framing enactments where scientific terminology and definitions are applicable, whether there would not be an advantage in the admission of the phraseology, particularly of naturalists, which embraces such exact definitions that they cannot be evaded or misinterpreted, and therefore would in that particular obviate all the vagueness so often reproached to the framers of legal instruments.

The unbreakable laws of science could thus help frame unbreakable laws. It was clear, Griffith noted, that the proper definition of a fish, such as that given "in the writings of modern naturalists"—"vertebrated animals with red blood breathing through the medium of water by means of branchiae"—would have headed off all of the contention in this notorious case, and prevented the public foolishness witnessed in New York.[64]

What is perhaps most interesting in all of these responses to the case

---

[64] The citations here and above are from Cuvier, *The Animal Kingdom, Arranged in Conformity with Its Organization, with Supplementary Additions to Each Order by Edward Griffith* (London: Whitaker and Co., 1827–1835), vol. 10 ("The Class Pisces," with additional notes by Charles Hamilton Smith), p. 27. The note appears in a section of this book written by Griffith, and the reference does not originate in the first (1817) or second (1829) French editions of the *Règne Animal*. My guess is that Griffith himself is the first source of news about the trial on the other side of the Atlantic, and that he learned of it through his association with the Academy of Natural Sciences in Philadelphia, which he served as a corresponding member. It is worth noting that English law tended to amplify the term "fish" by means of the expansive

of *Maurice v. Judd* is the strange erasure of what actually happened immediately after the much-ballyhooed verdict of December 31, 1818. For while the jury's rapid and sweeping verdict seemed to strike down both prongs of Judd's defense, declaring in a stroke that whales were fish and whale oil fish oil, the jury did not, in the end, have the last word on the matter. No sooner had the verdict been delivered than Judd sued for appeal, and the judge— informed that, in light of the controversy sparked by the case, the state legislature had indeed begun to move toward revising the statute—agreed to stay the final resolution of *Maurice v. Judd* pending clarification in Albany as to the proper scope of the "fish oil statute." The state representative from the city of New York, Peter Sharpe (who, recall, had testified in the trial, and endeavored to inform the jury that the true intent of the legislature extended only to liver oil from bona fide fish), immediately brought the law back into committee, and he succeeded in getting revised language to the full senate in the January term of 1819.[65] So, less than a month after the initial ruling in *Maurice v. Judd,* Sharpe brought to Albany an amendment of the original statute. He introduced the proposed changes on the 29th of January, and the Albany papers picked up the story on the 2nd of February, with the New York City papers following the next day, when the *Commercial Advertiser* listed the petition for amendment under its "reports on the Legislature of New York in assembly":

> Mr. Sharpe, from the committee to whom was refered [*sic*] the petition of sundry inhabitants of the city of New York, reported a bill authorizing the inspection of liver oil, and excepting sperm or whale oil from inspection—read and committed.[66]

The *Journal of the Assembly of the State of New-York* gave a synopsis of Sharpe's argument to the full house, which more or less followed the tes-

---

phrase "or creatures living in the sea," in order to spell out the full scope of legislation bearing on the whaling and fishing industries. See 49th Geo. III c. 98 § 37, and Scoresby, *An Account of the Arctic Regions,* vol. 2, "Abstract of Acts of Parliament," appendix I.

[65] Why was Sharpe so interested in all of this anyway? Imagine my surprise when I happened across a sequence of advertisements in early-nineteenth-century New York newspapers wherein one "Peter Sharpe" announced that he had for sale "1000 Umbrella and Parisoll [*sic*] Frames" at his shop on Maiden Lane—"where may be had at all times, *Whalebone* cut to any size or pattern" (*Public Advertiser,* 4 May 1807, p. 1). It would seem that Sharpe knew his whale products first hand! (It should perhaps be added that he also shared a position on the board of directors of a prominent New York charity with Mitchill and DeWitt Clinton; see the *Commercial Advertiser,* 29 July 1818, p. 2).

[66] For Albany report, see *Albany Argus,* 2 February 1819, p. 2.

timony of the city's chandlers and dealers in whale products. Sharpe explained that, having examined petitions and affidavits from knowledgeable parties, "it appears to your committee" that sperm and whale oil were "taken from the blubber, covering the bodies of the several species of whales, and extracted therefrom by fire, by the same method that tallow or lard is purified." Whale oil was thus "free from water and pure in its nature, and . . . no possible advantage can be derived to any one by having it inspected, except to the inspector." Estimating the direct costs of permitting such an inspection regime in the city, Sharpe calculated that the surcharge to consumers might reach as high as $600 per cargo of whale oil unloaded in New York or Sag Harbor, in a period when there was interest in stimulating the trade in whale products through New York ports. Given that half a dozen ships were returning each year to Sag Harbor from right-whaling voyages to the Brazil Banks (along with, now and again, an odd sperm whale voyage out of Hudson waters), the state could, in Sharpe's view, ill afford to impose such an impediment to "a business highly advantageous to our state and to the country at large." Such costly inspection of whale oils would "operate as a discouragement, if not a prohibition of the whaling business from the state of New-York" and needlessly raise the prices on "an article of great home consumption" as well as products very valuable to trade with the "East and West-Indies and Europe." After making this strong case against whale oil inspections, Sharpe rehearsed the valid arguments for the gauging and inspection of true fish oil, i.e., the messy and impure leached oil from the livers of groundfish. This unreliable product, used by curriers and tanners, did indeed demand legislative attention, and, as Sharpe put it, "public benefit will result from the inspection of oil of this description."[67]

On the 5th of February Sharpe's revised bill passed, and the following emendation was tacked on to the end of the "Act authorizing the appointment of gaugers and inspectors of fish oils":

> Be it enacted by the people of the State of New-York, represented in senate and assembly, that from and after the passing of this act, liver oil, commonly called fish oil, shall be inspected agreeably to the provisions of the act passed March 31, 1818, and hereby amended; and that all other oils shall be exempt from in-

[67] For the summary of Sharpe's case to the assembly, see New York (State) Legislature, *Journal of the Assembly of the State of New-York*, pp. 262–263. On the number of voyages in these years out of New York ports, see Starbuck, *History of the American Whale Fishery*.

spection, any thing in the act hereby amended to the contrary notwithstanding.[68]

In the end, whales, legally speaking, were not going to count as fish in the state of New York.

Not that this got Samuel Judd off the hook. On the 27th of February Richard Riker reconvened the court, and, perusing the new amendment, deployed an unimpeachably perverse legal analysis to *affirm* the jury's verdict against Judd: since the legislature had decided that the law needed amendment so as not to reach whale oil, he reasoned, this as much as *proved* that the previous language of the statute *did* reach whale oil; otherwise, why did it need to be *changed* so as *not* to include it?[69] Thus, the recent action of the legislature had made it clear that the version of the statute under which the action against Judd was originally brought did indeed call for the inspection of all marine oils, and therefore Samuel Judd owed James Maurice $75 for the fine, plus costs, which the jury had set at a token amount, but which Riker used his discretion to increase to a total of $72.27, bringing the total assessment to $147.27.[70]

It was, however, a small price to pay for what the trial had achieved: oil merchants and chandlers would never again pay for inspection on their spermaceti or whale oil. Disgusted with his newly circumscribed jurisdiction, James Maurice turned in his commission, sold his real estate in the city, and it appears that he and his wife moved out of town.[71]

[68] Cited in *IWF*, p. 79.

[69] Ibid.

[70] See the manuscript pleadings in the case, under "judgement." Old Records Room, County Clerk Archives, New York Supreme Court.

[71] Sampson mentions Maurice's resignation on 17 February (*IWF*, p. 79). For the sale of the Maurice property, see the manuscript "Deed from James Maurice and Jane his wife to John Flack, February 19, 1819" for two lots at Elm and White streets. Archive of the New-York Historical Society.

SEVEN Conclusion

NEW SCIENCE, NEW YORK, NEW NATION

Back in the introduction I gave three reasons for taking up a detailed study of *Maurice v. Judd* from the perspective of the history of science, and I want here by way of summary and conclusion to return to those three themes, reviewing them in the following order: first, *Maurice v. Judd* as an occasion to investigate cetaceans as "problems of knowledge" between 1750 and 1850; second, *Maurice v. Judd* as a window onto the contested terrain of zoological classification in this same period; and third, *Maurice v. Judd* as an opportunity to assess the broader place of natural history (and the sciences more generally) both in New York and in the United States as a whole in the early nineteenth century.

Turning to the first of these considerations: There can be, I think, little doubt that whales were "problematic" organisms on the watershed of the nineteenth century. The conflict at the heart of this study shows as much, and I suspect one would find similar complexity were one to consider the mythical, symbolic, and monstrous figurations of these creatures across a wide range of cultures and many centuries. Massive, furtive, valuable, at times dangerous, the whale has long occupied a place in the imagination of sea-conscious peoples proportional to its great bulk. Where mighty mysteries are concerned, systematic investigation is likely to be a fraught matter. Indeed (as Melville and Sampson both knew well), within the framework of a dutiful Christian natural theology, cetology—formal whale-knowledge—could even have about it the scent of the impious and the transgressive, particularly in view of the unique position accorded to "Leviathan" in the Book of Job. For there man's ignorance of the sea-giants stands as nothing less than the final and thunderous trump card played by God against the whole game of inordinate human intellectual aspirations: Job knew nothing of Leviathan; how then could Jehovah be expected to answer questions about the nature of evil?

It is one thing, however, to talk in a general way about whales as "mysterious" creatures, and yet another to dig in on a particular time and place with an eye toward understanding who knew what about these

animals, how, and in what ways they authorized their claims. It is this
latter project that has been my preoccupation throughout this book:
recognizing *Maurice v. Judd* as an arena where different kinds of whale-
knowledge would be permitted to stake out territory, evidence expertise,
and joust with opponents, I have used the Mayor's Court as the centering
point for an examination of four different communities whose claims
about whales were at issue in the trial—university-trained natural philosophers, practical whalemen, businesslike men of affairs, and "everyone
else," the ordinary English speakers of New York. This human taxonomy
followed the groupings offered by the actual participants in the trial, who
themselves recognized that the proper investigation of natural order was
contingent on a proper parsing of social class, guild, and geography. Separate chapters took up whale-knowledge among these different groups,
and marshaled diverse sources—natural history lecture notes, school
primers, biblical commentary, published taxonomic systems, manuscript
logbooks, sailors' doggerel, commodity market broadsheets, private correspondence, museum inventories, curriers' handbooks, etc.—in an effort
to recover and sketch the different ideas about whales and dolphins that
circulated in this period. Wherever possible I have tried to stress the dynamic quality of these different bodies of knowledge: literate whalemen
had their opinions about bookish cetology, and university zoologists
sounded whalers on the anatomy and behavior of their quarry; changing
prices and availability bore on the taxonomic hierarchies used by men
who traded in whale products, and who thus had their own ideas about
the relevant groupings of the animals from which those products came.
In each case it was possible to return to the court itself and to see these
different claims—and these different kinds of authority—set against
each other in the forensic confines of the trial, where each assertion had
to withstand assaults rhetorical, empirical, and even epistemological.
What a witness knew, how he knew it, and, finally, who he was were all
fair game in the Mayor's Court, and each of these distinct dimensions of
the testimony could be used to undermine the others.

Approached in this way, *Maurice v. Judd* has yielded a number of insights about the knowledge of cetaceans as it evolved during the hundred years between the publications of the tenth edition of Linnaeus's
Systema Naturae (1758) and Darwin's *Origin of Species* (1859). In chapter 2,
which worked to set up what ordinary New Yorkers considered the
"doxa" on whales and fish in 1818, a survey of public cabinets of natural
history in the city revealed the anomalous treatment of whale specimens
in the most popular collection open to curious citizens: in Scudder's

American Museum bones from the skull of a whale were catalogued with "miscellaneous" objects, a class that defied the seemingly basic distinction between the products of nature and the products of art; this placement can be seen to highlight the degree to which these animals resisted even the most fundamental classifications, since the exceptional character of whales was built into the very configuration of the city's new and highly visible temple to the sciences.

The presiding figure at that temple was Samuel Latham Mitchill, who figured as the star witness for the defense in *Maurice v. Judd*, where he was presented explicitly as the embodiment of science and reason. Chapter 3 took up Mitchill's efforts to install regard for natural history in the city in the first decades of the nineteenth century, his teaching and research in the sciences of classification, his particular interest in the sea, his specific knowledge of the cetes, and, finally, the fate of his trial testimony that "a whale is not a fish." I showed that Mitchill went to great lengths to teach natural history as a dynamic, inclusive, and democratic science, one that traded on the expansive energies of American sailors and merchants, and one that welcomed new arguments and original ideas. Turning to the trial, I contended that Mitchill's adversaries turned this vision of natural history against itself, stripping Mitchill's rhetoric of its populist trappings, and using his own emphasis on disputation and novelty to corrode the authority of his "new philosophy" and the "modern school" in taxonomy. By underlining that Mitchill's systematics carried implications for the place of human beings in nature, and by indulging in well-informed exercises of classificatory *reductio ad absurdum*, the lawyers for James Maurice succeeded in depicting Mitchill and his acolytes as risible at best, and quite possibly as something much worse—as slippery and protean purveyors of an amphibious philosophy that dissolved essential boundaries between men and beasts, and thus constituted, perhaps, a veritable threat to civic life.

At the heart of this charge lay Mitchill's shocking declaration that a whale was "no more a fish than a man." In a section bearing that title I situated this remark, showing that Mitchill—who had conducted considerable investigations in ichthyology—was acutely aware of the changing ways that systematists had endeavored over the previous fifty years to draw boundaries between fish and cetes. His insistence that whales and dolphins belonged in the Linnaean category of the Mammalia reflected a broader commitment to the radical implications of this class. In assessing Mitchill's efforts to settle the taxonomic status of the whales, I argued that the reception of his arguments must be understood in a wider cultural and intellectual context: generally speaking, this episode of "taxonomy at the bar" is part of a larger story of American responses to the rapid development of comparative anatomy in Paris and Edinburgh, and the whole affair certainly demonstrates the durability of Buffon in English book-trade natural history; at the same time, *Maurice v. Judd* reflects a public fascination with monsters quite specifically linked to very local preoccupations, for instance the "craze" surrounding the Gloucester sea serpent, and the contemporary unearthing of the bones of antediluvian mammoths.

Set against Mitchill's textual knowledge of whales (which he derived from the stack of Anglo-continental tomes he kept to hand when practicing natural history) was the testimony of whalemen who had encountered these creatures in full career in the open sea. That the two such men to testify in *Maurice v. Judd* did so on opposite sides of the question served as the point of departure for chapter 4, which revisited the problem of "the whaleman's natural history," and sought to reconstruct what whalemen understood about their prey, and into what broader ideas about nature that knowledge fit. Using both published sources and the clues scattered in whaling manuscripts, I offered a revisionist assessment of the traditional view that whalemen took little interest in natural history. On the contrary, the minute details of cutting-in diagrams bring into evidence a largely tacit vernacular anatomy and physiology of some sophistication. In addition, I worked to demonstrate that a broader natural history of the sea surface was implicit in a whaler's-eye-view of the whale and the ocean. If whalemen's knowledge of whales was intimate, it was by no means settled nor entirely settling, and the blubber-hunter's nomenclature translated imprecisely into the language of formal natural history: Did whalemen think whales were fish? Notwithstanding the writings of Hohman and other scholars of the whaling industry, most did, but many of them also expressed ambivalence about this ranking. Moreover, among whalers, not all whales were even, properly speaking, "whales"—a term many of them reserved for the whales that could be turned into dollars.

Finally, in chapter 5 I stalked "the whale in the Swamp," in order to show that, above all, *Maurice v. Judd* represented a showdown between powerful factions within the city's community of commercial elites. The dispute articulated itself in taxonomic terms, but the contours of the classifications traced boundaries between buyers and sellers, leathermen and chandlers, Knickerbockers and Yankees. At issue were changing economies of production, and lobbying efforts to create new legal sanctions and protections for the industrial ambitions of a rising generation of merchant-financiers. By tracing the implicit and explicit taxonomies of craft and the market, this chapter not only revealed the ways that money and labor gave shape to particular ideas about the order of nature, but also demonstrated that any effort to understand whale-knowledge in the nineteenth century demands a detailed investigation of these animals as commercial entities and must follow their course through the channels of trade and production. In many ways whales in this period loomed even larger in stacked casks than they did in the open sea.

So much for *Maurice v. Judd* as an occasion to consider whales as problems of knowledge between the 1750s and the 1850s. But what about the not-unrelated second rationale for so close an investigation of this trial: *Maurice v. Judd* as a window onto the contested terrain of zoological classification in this same period? Here I argued from the outset that this case would afford a valuable opportunity to contribute to a growing revisionist literature that has called into question a dominant narrative in the history of science—a narrative that considers the late eighteenth and early nineteenth century the "golden age" of the classificatory sciences, a period that saw the emergence of powerful new tools for arranging plants and animals into incontrovertibly coherent categories, categories that reflected the real affinities of these natural kinds and, by doing so, provided the conditions of possibility for the revolutionary genealogical and dynamic accounts of organic form and organization that would radically alter human understanding of nature. As I suggested in the introduction, recent scholarship in the history of science has shed new light on this story: by focusing on "difficult" organisms and adopting a perspective eccentric to the landmark publications in natural history—in short, by working under and around a familiar pageant of eloge and citation—a number of studies have demonstrated that the late Enlightenment victory of the classifiers, for all the strength of its metropolitan redoubts, did not extend very deep into the hinterland. On the contrary, communities of lay expertise—animal breeders, fanciers, veterinarians—had very different ideas about the order of nature, ideas grounded in particular kinds of close knowledge of particular taxons. From the perspective of these groups the early nineteenth century looked less like a taxonomic calm before the Darwinian storm (as one strand in the traditional historiography would have it), and more like an era of instability and change, when established orders were under siege and taxonomic expertise various and hotly contested. Nor were the perspectives of these "other" experts entirely overlooked by practitioners of the new classificatory sciences of comparative anatomy and systematic botany. By remaining sensitive to the migrations of ideas, terms, specimens, and practices across the quickening boundary between lay and "professional" naturalists, new histories of natural history become possible.

Maurice v. Judd has provided a rewarding point of entry for an investigation of these themes in the context of the United States. I am convinced that this examination of the case has demonstrated that, rightly understood, this period saw not the comfortable installation of a new and presiding natural order, but rather something much closer to an emerging "vacuum of zoological authority" in the early Republic.[1] Mitchill's testimony and the many-leveled rejection of his efforts provide compelling evidence on this point. But at the same time, I would like to offer this detailed treatment of *Maurice v. Judd* as something more than a transatlantic extension of an argument already advanced for Britain in the same period. After all, it is not enough simply to recover and rehabilitate largely forgotten "underworlds" of taxonomic activity and zoological expertise. Sure, there is something to be said for merely digging out these texts and voices, but such a program cannot disregard the important distinctions between these communities and the networks of increasingly formal natural history in the period. A more expansive view cannot be won simply by leveling the ground. On the contrary, by widening the stage on which we can see the history of natural order played out, we in fact create new opportunities to illuminate why certain players take center stage, steal the best lines, wind up as stars. In other words, while there is a value in showing that this period witnessed a much louder and more cacophonous conversation about the system of nature than the scholarship might have led us to expect, such a showing must serve as the point of departure for investigations of the emergence of the natural sciences themselves: in manifold ways this boisterous conversation was not just about animal groupings, it was fundamentally about a human taxonomy as well, since it is not merely that out of the hubbub came the categories of "whale" and "fish"; out of the hubbub also came the categories of "scientist" and "layman," "philosopher" and "fool." Hence, when at a dreadful family dinner in 1827 the eighteen-year-old whippersnapper Charles Darwin heard the pompous society physician Dr. Henry Holland declare whales to be cold-blooded, what the ambitious young naturalist actually heard him saying was, in effect, "I am a cad."[2]

[1] This is Ritvo's phrase, discussing Victorian Britain: Ritvo, *The Platypus and the Mermaid*, p. 14.
[2] The episode is recounted in Adrian Desmond and James R. Moore, *Darwin* (New York: Norton, 1994), p. 45. Darwin himself would later muse about the evolution of whales in the *Origin of Species* in an imaginative passage that became one of his most memorable "just so" sto-

With these problems in mind, I have made a concentrated effort throughout this study to show in detail how different orderings of nature served to delineate and ratify different social orderings: in the Mayor's Court whether whales were fish came to hang on matters of class, labor, and geography. That James Reeves was a sailor before the mast positioned his opinion with respect to that of Captain Fish in ways deemed significant to the resolution of the question. That Gideon Lee considered himself a "*northern* man" and not an "easterner" inflected his taxonomy in a meaningful way. Throughout the trial, taxonomies of nature defined communities of practice and figured social identities. It was in an extension of this same process, as I will argue below, that Mitchill's testimony on the question before the court gave important shape to the persona of the natural philosopher; in this way the case of *Maurice v. Judd* ultimately became a referendum not merely on a quirky classification of some odd beasts, but on the very place of science in the political life of the city and the state.

But before turning to that matter, what if we focus for just a moment on a more narrowly conceived history of the classificatory sciences: Does *Maurice v. Judd* have anything to offer here? I believe it does. In order to make sense of Mitchill's testimony, it proved necessary to investigate closely the process by which a set of marine creatures came to be codified, subdivided, and eventually redistributed across the emerging orthodox natural system of the early nineteenth century. I showed that this occurred later, and with less certainty (and more dissent) than has been commonly recognized, and that the moves traced—even defined—larger shifts in zoology in this period. What came to light was the durability of older groupings—"quadrupeds," "fish-like animals"—that were resurrected in these years to serve new functions in the face of some of the unnerving juxtapositions implicit (or even explicit) in the Linnaean sys-

---

ries about speciation: "In North America the black bear was seen by Hearne swimming for hours with a widely open mouth, thus catching, like a whale, insects in the water. Even in so extreme a case as this, if the supply of insects were constant, and if better adapted competitors did not already exist in the country, I can see no difficulty in a race of bears being rendered, by natural selection, more and more aquatic in their structure and habits, with larger and larger mouths, till a creature was produced as monstrous as a whale." Charles Darwin, *On the Origin of Species: A Facsimile of the First Edition* (Cambridge, MA: Harvard University Press, 1964 [1859]), p. 184. It is a passage that reminds us that "insect" retained well into the nineteenth century its etymologically primary meaning of "an animal divided *in*-to *sect*-ions," and thus remained a common term for the creatures we now call crustaceans. Nevertheless, it is worth noting that by 1859 Darwin could be thoroughly dismissive of the remaining whale-a-fish holdouts: "No one regards the external similarity of a mouse to a shrew, of a dugong to a whale, of a whale to a fish, as of any importance" (*On the Origin of Species*, p. 414).

tem, and in its at times disconcerting elaboration by the practitioners of comparative anatomy. The movement into the interior of organisms, and the commitment to the idea that hidden anatomy would instruct adepts in the differences-that-made-the-most-difference in nature— those two pillars of the "new philosophy"—were revealed to be shakier, and harder to transport and erect outside the precincts of the *Jardin* of the Paris Museum of Natural History than might have been expected. Just how much difference *did* deep internal anatomy make in the consideration of an animal whose whole social reality in the city of New York could be more or less reduced to an external layer eight inches thick? If whalemen testified that the whale sometimes vanished into the depths for indefinite stretches, or that it blew water in its spout, then could the mere fact of its having lungs be taken to prove that it could not breathe underwater? The case of *Maurice v. Judd* thus suggests that not enough attention has been paid to resistance to Cuvier's scalpel as the hegemonic instrument in the classificatory sciences of zoology. As the work of Blainville and Swainson later in the century would demonstrate, there remained those who insisted that external appearances held vital clues to natural order: for Swainson, the superficial similarities between whales and fish became an essential linkage in the geometry of the quinary system; for Blainville, there was something manifestly ridiculous about a system for understanding living creatures that was powerless until it had a dead body on the dissecting table. These were both views, one suspects, that a Nantucket whaleman would have appreciated.

At the same time, this discussion of *Maurice v. Judd* has suggested the importance of restoring the centrality of the tripartite organization of biblical zoology to the taxonomic debates of the late eighteenth and early nineteenth centuries.[3] Updating his father by letter after the trial, the young Henry Meigs wrote that the whole whale/fish affair had put him in mind of a couplet in Samuel Butler's mocking *Hudibras* (circa 1662) where the "old philosophers" came under scrutiny for their preposterous "grand distinctions." As Meigs put it, such quixotic systematists of antiquity famously "included ships and all other things in or on water, among the *Fishes!* Windmills &c. on land for *Animals Terrestrial!* And flying arrows or wigs, [as] *birds!*"[4] A careened ship, by these lights, ceased

---

[3] See Ritvo, *The Platypus and the Mermaid*, p. 46, for a discussion of this issue.

[4] H. Meigs to J. Meigs, 2 January 1819, Meigs Papers, New-York Historical Society; emphases in original. The line from *Hudibras*, which Meigs (mis)cites in his letter, is from canto 2 of part 2, "That Bonum is an animal, made good with stout polemic brawl." The reference is to the debates among the Stoics over the nature of the good. Butler's own footnote expands the

to be a fish, since it had been pulled ashore. Meigs's erudite and scathing commentary on popular sensibilities is yet further evidence that the Genesical division of animals into those that fly, those that swim, and those that creep was pervasive and tenacious, and that a self-consciously vanguard thinker like Meigs saw it as the fundamental adversary to Mitchill's modern taxonomy. This observation and much evidence in the trial transcripts indicate just how entrenched this vision of nature really was, and suggest that it has not been given adequate attention by most histories of natural history. It would be worth revisiting the emergence and reception of synthetic taxonomic systems in the eighteenth century with an eye toward understanding how that threefold architecture endured, came under siege, or demanded to be reckoned with. If historians of science interested in natural history want a "Copernican Revolution" around which to bend the narrative of their subject (though I am by no means sure such a thing is needed), the story of the breaking of the Genesical triangle might well serve that function. Whether such an investigation would bear fruit or not, there can be little doubt that the basic divisions staked out by that older system undergo a transformation and appropriation over the period 1750–1850. So, for instance, perhaps the very best way to understand the emergence and resolution of the question "Is a whale a fish?" is less as the history of where to slot a big, strange, and very unusual animal, and more as the history of the appropriation of the word "fish" as a scientific term; indeed, it was only possible to *ask* this "paradoxical" question when a familiar and humble word was obliged to do duty in an increasingly abstruse and systematic scheme for the natural order.

When William Sampson set down his "accurate report of the case of James Maurice against Samuel Judd," he chose as the epigram for the pamphlet a shining line penned by his fellow Irishman, the politically active playwright Richard Brinsley Sheridan: "Who says a whale's a bird?" The line appears in Sheridan's self-conscious 1779 comedy on the world of the theater, *The Critic,* where it features in an ultra-dreadful pastiche of Ophelia's mad scene staged helter-skelter for the benefit of a pair of preening critics named "Puff" and "Sneer."[5] It took a character of Sampson's intellectual breadth (and wicked wit) to dig up so apposite a refer-

---

joke in the direction that interests Meigs: "The same authors are of the opinion, that all ships are fishes while they are afloat; but when they are run on ground, & laid up, in the dock, become ships again."

[5] Richard Brinsley Sheridan, *The Critic,* edited by David Crane (London: A. and C. Black, 1989), p. 81.

ence, since he had in his sights not merely the general goofiness of Mitchill's "new philosophy," but a very specific element of his taxonomic system. It turns out that among the distinctive revisions Mitchill wrought upon the Linnaean/Cuvierian classifications that he taught at the College of Physicians and Surgeons was an original effort to smooth the apparent discontinuity between his penultimate zoological class, *Aves,* and the class with which his cycle of lectures climaxed (and terminated), the Mammalia. To this end, Mitchill reconfigured the orders within these adjacent classes so as to place the order of waterbirds (like ducks and coots) last among the *Aves,* and the order of cetes first among the mammals. The result was, in his view, a pleasing continuity across the taxonomic divide: water-birds took up their place beside the water-beasts, in the greater service of a clear and coherent *scala naturae* rising smoothly through creation. When read in the context of this fact about Mitchill's natural history, Sampson's quip about whale-birds (a quip taken up, as we have seen, by others) must be understood as a dart precisely aimed at a chink in the armor of Mitchill's system. Because coots and other waterbirds had, notoriously, been treated as fish for the purposes of early modern Catholic regulations for the consumption of meat on Fridays, the notion of a bird-fish already belonged in an established tradition of Protestant mockery of benighted scholastic obscurantism.

This point is worth raising not merely because it brings the last of the trial's inside jokes into the open, but more importantly for what that joke tells us about problems seen to be central to early-nineteenth-century classification. Anomalous organisms—whales, bats, coots, monkeys, and even human beings—could function both as *exceptions* to defined groupings, and as *links* between them; the tension between those treatments exposed vulnerable sutures in the surgical precision of natural history in this period. It would be interesting to understand better how this tension was handled, and to trace how the meanings of continuity and exception changed in the period. Given the significance of these ideas to Darwin's work at mid-century, there is every reason to think that such an investigation would shed considerable light on later developments.

In these and other ways *Maurice v. Judd* has provided a sidelong, but, I believe, idiosyncratically panoramic view across the combative field of zoological classification in the period 1750–1850, and has suggested several directions for further work.

Finally, then, what can be made of this case under the third and last rationale for the present study: *Maurice v. Judd* as an opportunity to assess the place of natural history (and the sciences more generally) both in New York and in the United States as a whole in the early nineteenth century? On the narrower question of the case in the context of the intellectual culture of the city in this period, I have already alluded to the significance of the timing and urban geography of the trial, and underlined the stature of the man called upon to represent the views of the sciences to the jury: *Maurice v. Judd* took place in the Mayor's Court in the new City Hall, less than 100 yards from the newly founded New-York Institution of Learned and Scientific Establishments. A member of nearly all those establishments, Mitchill could not but represent this Clintonian "Royal Society," which, under public patronage, had just brought together at the civic center of the city clubs of well-heeled virtuosi and emergent specialized scientific societies like Mitchill's Lyceum of Natural History. All of these savants conducted their affairs under the banner of the well-advertised pantheon of science and culture, Scudder's American Museum, which was right upstairs (and had prominent signage on the façade of the Old Almshouse building).

In his study of "The Role of De Witt Clinton and the Municipal Government in the Development of Cultural Organizations in New York City, 1803–1817," Kenneth Nodyne uses the founding of the New-York Institution to argue that 1817 represented an *annus mirabilis* in the cultural and intellectual life of the city.[6] As the city's dispersed learned and scientific communities took up residence in the nascent Institution, and as Governor Clinton—flush with the promise of the Erie Canal that he was steering to legislative ratification—presided there (as "His Excellency") over extravagantly formal meetings of the Literary and Philosophical Society, and as Mitchill's energetic new Lyceum of Natural History came into being and staked out its turf in the same edifice, there seemed to be reason to celebrate the dawn of a new era in the relationship between science and society in the state of New York.[7] But, as Nodyne and others have shown, the optimism was to be short-lived. Clinton, Pintard, and their allies were, in the end, unable to secure the passage of "an Act for the encouragement of the New-York Institution for the promotion of the Arts and Sciences" through the Albany assembly in 1818–1819; there would be no ten-year excise tax on tavern licenses to underwrite

---

[6]Nodyne, "The Role of De Witt Clinton." This dissertation has served as the most important source for most of the other secondary treatments of this episode.

[7]On DeWitt Clinton's (much disputed) scientific identity, see: Vivian Hopkins, "The Empire State: De Witt Clinton's Laboratory," *New-York Historical Society Quarterly* 59, no. 1 (1975), pp. 7–44; and Jonathan Harris, "De Witt Clinton as Naturalist," *New-York Historical Society Quarterly* 56, no. 4 (October 1972), pp. 265–284.

the activities of minerology and zoology at the Old Almshouse. While a host of factors conspired to delay and finally sink the bill (including the highly personal animosity of the anti-Clintonians—particularly the virulent energies of Gulian C. Verplanck), a rising tone of general dismay met the New-York Institution in 1818 and 1819, reinforced by increasingly loud echoes of some of the early complaints about the whole project. From the start some New Yorkers had expressed irritation that the dilapidated almshouse had not been leveled, and the valuable lots sold off for public benefit. In the process Warren Street could have been opened to through-traffic, relieving the congestion around the political heart of the city. Others had pointed to the fact that, spruced up and converted into offices, the same building could have afforded the city thousands of dollars of income on a regular basis.[8] Support for a "Royal Society" in New York clearly met considerable popular resistance from a population steeped in Swiftian satires on gouty Sir Joseph Banks and his clownish coterie of FRSs who primped and preened in the decadent London original of Clinton's Gotham ambitions.[9] Who would choose to reproduce such an establishment in the young Republic? Why? In March of 1818, the *National Advocate for the Country* complained explicitly of the clique of prominent citizens—saliently Mitchill—who seemed to be intent on presiding over a publicly funded playground for self-styled philosophers, and throughout 1819 the New York *American* repeatedly took shots at the pretensions of Clinton and his gaggle of naturalists, going so far as to publish a mocking transcript of a "philosophical" discussion at the Institution, in which Mitchill's taxonomic preoccupation with "tails" was the occasion of much low humor.[10]

By 1819, in short, the outpouring of derisive scorn for the Institution, Mitchill, and Clinton had reached scandalous proportions. In that year Fitz-Greene Halleck would publish his long poem about the social life

[8] For a summary of these positions, see "New-York Institution," *American Monthly Magazine and Critical Review*, 1, no. 4 (1817), pp. 271–273, at p. 273.
[9] Such satires were well known. See, for instance, Gulian C. Verplanck, *The State Triumvirate, A Political Tale* (New York: For the Author, 1819).
[10] It would seem that this was a double joke, playing both on Mitchill's concern about the "Bucktails" (the anti-Clintonian political faction) and his odd ideas about zoology. See "A Literary and Scientific Conversation at the Hall of the Philosophical Society of Gotham," *American*, 23 November 1819, p. 2. See also "To the Author of Dick Shift," *American*, 30 October 1819, p. 2, which jeered the juxtaposition of Clinton and Newton reportedly on display in the portrait gallery of the New-York Institution. (NB: The *American* began publishing in March of 1819.) For the complaints of the *National Advocate for the Country*, see Nodyne, "The Role of De Witt Clinton," p. 176.

of New York, "Fanny," which would take its swipes at Mitchill and the other philosophers-on-the-dole in the city. For instance, Halleck called for everyone to

> Bless the hour the Corporation took it
> > Into their heads to give the rich in brains
> The worn-out mansion of the poor in pocket,
> > Once the "Old Almshouse," now a school of wisdom
> Sacred to Scudder's shells and Dr. Griscom.[11]

And Halleck went on to promise, tauntingly, that all the ancient philosophers of Athens would have a thing or two to learn from a visit to the virtuosi of New York:

> In short, in every thing we far outshine them—
> > Art, science, taste, talent; and a stroll
> Thro' this enlightened city would refine'em
> > More than ten year's hard study of the whole
> Their genius has produced of rich and rare—
> > God bless the Corporation and the Mayor![12]

Just how deeply such paeans were steeped in sarcasm is made clear by more vicious commentaries of the Croaker in the same period: for instance, the ode "To Quackery" (dedicated to "thou patroness of knaves and fools"), which complained of "charlatans" like Hosack, Mitchill, and Clinton, who continuously conspired to brew up "some strange melange / Of scientific Salmagundi." The poem went on to picture an unholy alliance of politician-naturalist and his philosopher-muse:

> Clinton! The name my fancy fires,
> I see him with a sages [sic] look,
> Exhausting nature and whole quires
> Of foolscap—in his wondrous book!
> Columbia's genius hovers o'er him
> Fair science, smiling, lingers near,
> Encyclopedias lie before him,
> And Mitchill whispers in his ear.[13]

And the viciously anti-Clinton poem by the Croaker from about the same time, entitled "The Modern Hydra" (which, like "To Quackery,"

---

[11] Fitz-Greene Halleck, *Fanny* (New York: C. Wiley, 1819), section 68.
[12] Ibid., section 48.
[13] This poem appears in Drake, *The Croakers*, pp. 129–131.

was never published, but circulated in manuscript), lambasted the governor for transmuting his stymied political ambitions into scientific showmanship, since, no sooner had his enemies "tomahawk'd his head political" than "straight from the bleeding trunk out slid his / well fill'd noddle scientifical."[14] An exercise of Herculean cauterization was emphatically the city's only hope.

In a way, this was exactly what happened. By 1827 public support for the New-York Institution would begin to be withdrawn, and Clinton's death the next year further eroded its constituency. By 1831, the year of Mitchill's passing, the Old Almshouse had been cleared of its freeloading philosophical tenants and the building converted into administrative offices servicing the municipality. Several of the constituent societies failed, and Scudder's Museum, divorced from the matrix of Mitchill's Lyceum and the other institutions of natural history, began its slip into the carnival clutches of P. T. Barnum. Confronting this failure of New York's first formal, public, and centralized institution of intellectual culture, Thomas Bender points to "[t]he failure of elite and fashionable society to justify its intellectual pretensions," and suggests that the episode must be read in the larger context of the growing cultural divisions in the city in this period. Rather than seeing the collapse as simply a retrograde step in the progress of science in the United States, it is necessary to recognize that the inviability of the New-York Institution reflected the emerging cultural and intellectual ambitions of a rising community of artisans and mechanics, who were seeking support for their own institutions for the advancement of learning (Gideon Lee, interestingly, was a leading figure in this movement). Understood this way, the New-York Institution was simply a model for intellectual culture that had passed its sell by date, peopled with a patrician class whose sense of entitlement no longer turned the key to the public coffers. By these lights, that the Lyceum survived, morphing over the years into the New York Academy of Sciences, stands as proof that the right kind of institution— more professional, more specialized, not aspiring to embody the *nous* of Gotham—could indeed endure.[15]

To be sure, more than one tremor shook the New-York Institution in the uneasy decade of its active life; everything from the Panic of 1819 to the founding of the New York Mechanic and Scientific Institution and

[14] Ibid., pp. 112–113.

[15] I am here summarizing the arguments made by Bender in chapter 2, "Patricians and Artisans," of *New York Intellect*, particularly pp. 76–81. See also Cornog, *The Birth of Empire*, pp. 65–69.

its mouthpiece, the *Mechanics' Gazette*, belong on the autopsy report. Moreover, there can be no doubt that the fate of the philosophers of the Old Almshouse was tightly tied to the fortunes of DeWitt Clinton, whose political trajectory declined through the 1820s.[16] But I believe the discussion of *Maurice v. Judd* in this book offers a uniquely early and charged glimpse of the dynamics that would usher the New-York Institution off the green of City Hall Park and into the annals of the (itself suddenly homeless) New-York Historical Society.[17] Bender refers to the mass eviction as the "first formal challenge to elite cultural hegemony" in Manhattan; rightly understood, *Maurice v. Judd* offers a premonition, even the first salvo of that challenge, since the case saw the Institution's familiar figurehead take the short walk across the park to visit City Hall as an expert, and return as the laughingstock of the island. In between, he had made no friends among a powerful group of "men of affairs" against whose claims he had testified (some of whom would lead in the creation of parallel institutions of learning that defined themselves in opposition to Mitchill, his kind, and their societies), and he had permitted an irrepressible lawyer-populist, William Sampson, to make a Punch and Judy show of the whole science of natural history, to the delight of the artisans, seamen, and other working people who crowded the gallery of the court. In the showdown between expertise and eloquence, eloquence drew fast and fired straight.[18]

So palpable was this shift—from triumphant entrance to ignomin-

[16] On the Mechanic and Scientific Institution (of which Gideon Lee was founding vice-president), see: Wilentz, *Chants Democratic*, pp. 40–42; Bender, *New York Intellect*, pp. 78–88.

[17] In this sense, *Maurice v. Judd* merits consideration in relation to the themes raised in Sven Dierig, Jens Lachmund, and Andrew Mendelsohn, eds., *Science and the City. Osiris*, 2nd series, 18 (2003). See also Arnold Thackray, "Natural Knowledge in Cultural Context: The Manchester Model," *American Historical Review* 79, no. 3 (1974), pp. 672–709.

[18] Or was it, to be more precise, a showdown between eloquent expertise (Mitchill) and an expert at eloquence (Sampson)? I was struck by the comment of one reader of this manuscript, Anthony Grafton, who saw in the exchanges of *Maurice v. Judd* a rhetorical contest deeply informed by the classical tradition (then entering its last years of undisputed dominance within university culture in the United States). This observation emphasizes the extent to which Mitchill and Sampson shared a formation in the Greek and Roman arts of disputation, and deployed those skills in the course of the trial right along with their knowledge of ichthyology. More work on the relationship between this sort of civic humanism (fundamentally at odds in many respects with the very idea of expertise) and the emerging ideal of the scientific expert would certainly shed additional light on the development of the sciences in the early Republic. For a superb introduction to the culture of oratory and debate in American higher education in this period, see James McLachlan, "The Choice of Hercules: American Student Societies in the Early 19th Century," in *The University in Society*, edited by Lawrence Stone (Princeton: Princeton University Press, 1974), vol. 1, pp. 449–494. On classical rhetoric in the early Republic, see Caroline Winterer, *The Culture of Classicism: Ancient Greece and Rome in American Intellectual Life, 1780–1910* (Baltimore: Johns Hopkins University Press, 2002), chapters 1 and 2.

ious retreat—that Judd's own lawyers apparently realigned their defense after Mitchill's departure. Sensing that the doctor's testimony had not won the day (and in fact might have done damage to their cause), Robert Bogardus began his closing arguments in the trial with nothing less than a disavowal of the whole episode: "gentlemen," Bogardus announced to the jury concerning the illustrious naturalist, "I did not call him as one upon whose testimony I meant to place the defence of my client, but rather, because it was proposed as a matter of amusement, and I consented, considering that it would be interesting to curiosity to hear what he would say upon a question of so much novelty in a court of justice." Such disputations among the worthy over trivial questions of natural history might make suitable entertainment for the philanthropic *conversazione* of the New-York Forum, Bogardus suggested dismissively, but this sort of chat had no place in the grave chambers of City Hall. In brief, Bogardus himself showed natural-historical polemic the door, announcing to the jury, "it will be well understood by you, as it is by the judge who presides, that juries have nothing to do with questions of science, which can only serve to perplex them; *much less are the theories of scientific men to be taken as the law of the land,* seeing how little philosophers are agreed amongst themselves." Jumping on the bandwagon of Sampson's subversive cross-examination, Bogardus went on to declare that men of science ought really play no role in the courts at all, since "you could never bring twelve of them, if empanneled [*sic*] as a jury, to be unanimous on any one speculative point." He then launched into a colorful taxonomic *reductio* of his own, asking the jury whether, if they were asked to determine the scope of an ordinance to inspect "fowl feathers," they would—on the strength of scientific testimony that an ostrich is rightly placed in the genus "fowl"—permit a hyperactive inspector to stand on guard at military parades, studiously inspecting the plumes of the generals and the ornaments in the ladies' hats.[19] Cheap laughs at the expense of bookish systematics had, by the trial's end, crossed the aisle.

This strategic about-face offers striking evidence of just how badly Mitchill and science were seen to fare on the stand; and while it is

---

More generally, Thomas Bender, ed., *The University and the City: From Medieval Origins to the Present* (New York: Oxford University Press, 1988), particularly chapter 9, by Louise L. Stevenson ("Preparing for Public Life: The Collegiate Students at New York University, 1832–1881").

[19]For the quotes in this paragraph, see *IWF,* pp. 54 and 56; emphasis mine. The New-York Forum was a discussion group organized by J. H. Hatch, L. Clark, and R. I. Wells, which took on questions like "Are permanent charitable institutions except for the purposes of education beneficial to society?" (debated on 29 March 1819); tickets were sold, and the proceeds benefited, among others, the New-York Sunday School Union Society. The group generally met at the City Hotel on Monday evenings. See *American,* 27 March 1819, p. 2.

possible to see Bogardus's repositioning as merely a tactical maneuver in a two-pronged defense offered by Judd's lawyers at the close of the trial (with Price making a closing statement that appealed to the jury on behalf of Mitchill's stature, and Bogardus providing an analysis in the alternative aimed at those jurors unpersuaded by the doctor's natural history), it is impossible to deny that Judd's lawyers decided they needed to distance themselves and their case from the testimony of the "Nestor of American Science." In the process, and in the resulting verdict of *Maurice v. Judd*, the citizens of New York distanced the Mayor's Court from the New-York Institution, and fended off what had come to be seen as the encroachment of the philosophers encamped at the gates of City Hall. To declare the whale a fish in 1818 was thus to express serious misgivings about the Clintonian vision of philosopher-kings, installed at the heart of a new Athens on the Hudson.

For clear evidence that such large issues were indeed at play, we need look no further than the final note Sampson appended to his own account of the trial. There, in an incandescent Swiftian parody of Clinton's Alexandrian projecting (and Mitchill's Stagerite service to power), Sampson offered a "modest proposal" to the patron of the Erie Canal and the Surgeon General of his militia. Since it appeared that mammalian whales and dolphins were, in the acceptation of philosophers, near cousins to man, it was only a matter of time—and proper initiative—to transform New York into the capital of a cetacean nation. Here was an eminently suitable task for the constituent societies of the New-York Institution:

> When it is considered that our waters abound with these dolphins, so inclined by nature to aid and succour us, that the larger kinds are only banished by our cruelty from our shores; and seeing it is well attested that their milk resembles that of cows, with the addition of cream, (see Dr. Brewster's Cyclop. article Cetology) would it not be worthy the wisdom of our statistical, agricultural, and oeconomical societies, to turn their attention to this weighty consideration, whether these creatures might not by good treatment be induced to lend their aid to the navigation of our waters, and to furnish us in return for our hospitality with abundance of nutrition?[20]

There was the matter of training these cetes to negotiate fresh water, so they may be "used in our great canals," but if the *canals* (wink wink) were dug deep enough and configured just right,

---

[20] *IWF*, p. 83; question mark added.

there is no reason . . . why judges and lawyers, legislators and politicians, office hunters, and lobby-members, may not, before many years, in their attendance upon the terms, enjoy the advantage of a conveyance upon a whale's back, infinitely surpassing the speed of the steam boat; and the shores of the Wallabout may resound with the music which calls the dolphins to be milked, and be studded with villas where the citizens shall repair to enjoy country air and dolphins' whey.

And this Georgic depiction of classical splendor on the banks of the Erie Canal could be extended to the heart of the city itself, since "The bay of Goannes seems designed by nature for the reception of the whales; from them will be derived a rich supply of butter and cheese for home consumption and foreign commerce." Moreover (here mocking Mitchill's role in the fortification of New York harbor during the preparations for the War of 1812),

Another important acquisition will be the defence and safety of our harbour. If one of Claudius's gallies was swamped by a single whale, overpowered and stranded as it was, what would the fire of a three decker, or a phlogobombos, avail against an inundation from the snouts of three hundred well disciplined whales? It is evident that the use of fire engines will be superseded.[21]

Sampson's fantastic vision—of the philosophers' whale, trained up by the good ministrations of energetic virtuosi, doing service in the Clintonian canals, while attending the expansion of the Empire State both upstate and *outre mer* conjures a Hogarthian tableau of *Maurice v. Judd*, and situates it firmly in the broader machinations of New York politics in the period. For all its good fun, this send-up of the eccentric energies of agrarian improvers, nature-nationalists, and society macaronis riding the whales to market makes it abundantly clear that through the trial flowed the strong currents of opposition to the institutions, innovations, and schemes of state-sponsored "philosophy." Science in the service of the state looked to many New Yorkers suspiciously like the state in the service of the men of science, and while the New-York Institution still shared its triangular lot with City Hall, there would be less traffic across the green in the years to come.

By the summer of 1819, then, when *Maurice v. Judd* entered into urban legend, the *annus mirabilis* of New York learned culture was well and truly ended.

---

[21] Ibid; question mark added.

If locating *Maurice v. Judd* in the local geography of New York City permits us to see the ways that the trial was both embedded in, and helped shape, wider debates about science and society in Manhattan, does this study of the case have anything to offer to a still broader consideration of such debates in the early Republic? For starters, we might merely note its surprising reach as "news" in the world of reading Americans at the start of the nineteenth century: not only have I found (without particularly assiduous effort) reports of and allusions to the whale-a-fish case and its aftermath in the newspapers of fifteen of the twenty states that comprised the Union at the time of the trial, I have also turned up additional references in the many papers of Washington, DC, the backwoods presses of Maine (not yet separated from "Old Massachusetts") and West Virginia (not yet separated from the Commonwealth), and even the irregular columns printed in the remote reaches of the Arkansas Territory.[22] For a nation still mostly without a "national culture" this is surprisingly national coverage. Quirky debates over French systematics apparently made good copy in these years, suggesting broad curiosity about the order of nature in the young United States.

Moreover, while extrapolations from so peculiar an incident must be hedged about with qualifications, I believe, as I suggested in the introduction, that this detailed discussion of *Maurice v. Judd* invites us to raise a warning flag beside a dominant theme in the extensive literature that treats the larger meanings and uses of natural history in the United States across the first half of the nineteenth century. At issue is the preoccupation of a considerable number of commentators with the role of natural history collecting, taxonomy, nomenclature, and the display of *naturalia* in the "constitution" of what Perry Miller called "nature's nation."[23] For Paul Semonin, Christopher Looby, David Scofield Wilson, and others, a reading of the works of Jefferson, the Peales, the Bartrams, Catesby, and the whole canon of American authors attentive to the flora and fauna of the continent provides strong evidence for a distinctively "American" use of natural history in the codification, delineation, and articulation of national ambitions. From the practical to the symbolic, from ornithology to paleontology, from Lewis and Clark to Cole's *The*

---

[22]Much of this was, of course, reprinting of stories from other papers. An exhaustive list is impossible, but here are a few choice references: *Winyaw Intelligencer* (South Carolina), 30 January 1820, p. 4; *Kentucky Reporter,* 11 October 1820, p. 1; *Arkansas Gazette,* 9 September 1820, p. 3; *Louisiana Advertiser,* 27 July 1820, p. 1; *Farmer's Repository* (West Virginia), 20 January 1819, p. 3; *Maine Intelligencer,* 3 November 1820, p. 52.

[23]The foundational work in this area was William Martin Smallwood, *Natural History and the American Mind* (New York: Columbia University Press, 1941).

Course of Empire, we are reminded that Americans forged a particular kind of link between natural history and nationalism, and used the language of nature to narrate their nation, to call it into being.[24] For Wilson, an intangible "American spirit" whispers in such texts: he hears Jonathan Carver speak a language tellingly unlike that of Captain Cook, and senses in James Bartram's attention to his vegetables a "fellow human presence lending continuity to American culture as it stretches from the seventeenth century to the nineteenth century, and the twentieth."[25] Laura Rigal zeros in on taxonomy itself, and imaginatively figures the ordinary American at the dawn of the Republic as a kind of new Adam, a copy of Linnaeus at the ready: "With the *Systema Naturae* in hand, any provincial farmer or mechanic had access to a global organizational system through which all plants (and, eventually, insects, birds, and quadrupeds) emerged into visibility and knowability, precisely as if, at the moment of their creation, they had been stamped with a species identity by some original Author of all being."[26] It could have been like that, except, as we have now seen, it wasn't like that at all: the laboring folk we have met in this book thought Linnaeus mostly absurd, and they had their own ideas about how to organize the natural world, ideas derived from labor, craft, scripture, and experience.

Nevertheless, for all the extravagance of these kinds of claims (and the historical misprisions they not infrequently reflect), it cannot be denied that late-eighteenth- and early-nineteenth-century American naturalists often justified their activities in patriotic language, and many of them—including Mitchill, as we have seen—expressed genuine enthusiasm for using the scientific study of nature to bring profit and autonomy to American enterprises; in the process, American science itself

[24]The literature here is immense, but a good point of departure is the classic work of Perry Miller, *Nature's Nation* (Cambridge, MA: Harvard University Press, 1967). While the broad theme has certainly been nuanced by more recent historical work (see in particular Joyce E. Chaplin, "Nature and Nation: Natural History in Context," in *Stuffing Birds, Pressing Plants, Shaping Knowledge*, edited by Sue Ann Prince [Philadelphia: American Philosophical Society, 2003], pp. 74–93), it nevertheless remains a somewhat preformatted thesis available to students in American studies, art history, and the historically oriented literary analysis of American texts. For a sense of where the field has been (and is) on these questions, consider: Semonin, "Nature's Nation"; idem, *American Monster;* Ann Shelby Blum, *Picturing Nature: American Nineteenth-Century Zoological Illustration* (Princeton: Princeton University Press, 1993); Pamela Regis, *Describing Early America: Bartram, Jefferson, Crèvecoeur, and the Rhetoric of Natural History* (DeKalb: Northern Illinois University Press, 1992); and Wilson, *In the Presence of Nature*.

[25]Wilson, *In the Presence of Nature*, pp. 86 and 122.

[26]Rigal, *The American Manufactory*, p. 5. While I am here singling out a hyperbolic moment in her text, it should be emphasized that this is a book from which there are plenty of things to be learned.

would grow increasingly independent, something Mitchill regularly declared a national desideratum.[27] In addition, it seems very likely that, as Looby argues in "The Constitution of Nature," natural history played a privileged role in the early Republic at least in part because it provided a powerful tool with which to confront the greatest peril in the formative years of an American nationality: "the threat of social disintegration that was posed by the conflicting racial, ethnic, religious, linguistic, sectional, local, and ideological categories of self-definition and social loyalty that were inhibiting the formation of a genuinely national identity."[28] A properly conceived and executed natural history of American nature might thus afford American citizens the shared "universal language" they so needed, and provide an integrative inventory of the new nation: an expansive and legible catalogue of a continent of particulars; a centrifugal enumeration always held in focus by the centripetal forces of word and mind.

Appealing as this account may be, and whatever truth it may capture, this foray into the cacophonous theater of *Maurice v. Judd* has I hope been a salutary reminder that the language of natural history was not impervious to the tug of disintegrative interests. On the contrary, as we have seen, the language of taxonomy could be used without difficulty to articulate sectarian division, and the nomenclature of nature could be held both to secrete and to unmask conflicting identities: in the end, the Yankee whale had no business in a Knickerbocker court. Calling nature's nation into being with the incantations of natural history could be, when the curtain fell, a very uncertain affair.

---

[27] Philip Pauly, who is currently at work on a book (on scientific definitions of "native" American flora and fauna) that will shed light on this issue, has already touched on these themes in the first chapter of *Biologists and the Promise of American Life,* see especially pp. 19–22.

[28] Looby, "The Constitution of Nature," pp. 259–260.

EPILOGUE: WHALES AND FISH, PHILOSOPHERS AND HISTORIANS, SCIENCE AND SOCIETY

Philosophy, in her turn, studies these beings. She goes both less deep and deeper than science. She does not dissect them; rather, she meditates upon them. Where the scalpel has been at work, she sets in with the hypothesis . . .
VICTOR HUGO, *Les Travailleurs de la Mer*[29]

Are whales fish? Today every schoolchild knows the answer: of course not. For this reason most modern readers, shuffling the dusty documentation of *Maurice v. Judd,* would be sorely tempted to see the case as a conflict between right and wrong, sincerity and expediency, hero and villain. While this is very much the story I have worked *not* to tell, it cannot be denied that Mitchill has about him the promising air of the scientific martyr, since he stood up for the truth in the face of snickering opposition. By these lights, insouciant Sampson, while likable enough, must play the role of the irresponsible sophist (some might even like to make him the patron saint of strong-program constructivists and their radical science-studies epigones). He attacked science and he won, and the costs turned out to be higher than he probably expected. Should he have known better? Perhaps. Maybe it all seemed like good fun when it was merely about ichthy-whatever. When, in the end, it turned out to be about the status of science itself in New York, about the proper relationship between natural history and social order in the young Republic, did Sampson feel a pang of regret for the damage that had been done? Perhaps. That orange file fish forwarded to Mitchill in the autumn of 1819 may indeed have been freighted with remorse.

It must be said, though, that the drama Sampson staged—the one that pitches society against science (or vice versa)—has been written and rewritten over the last five hundred years, and it always seems to draw a crowd: from Galileo's travails to those of Lysenko, from Scopes to intelligent design. For some, retelling these stories provides the occasion for parables about how societies must become "more scientific" (the people must be better educated, scientists must be granted a greater role in political life); told differently, the same stories become for others warnings

---

[29] Victor Hugo, *Les Travailleurs de la Mer* (Paris: Gallimard, 1980 [1866]), p. 440.

against insidious technocracies and scientific megalomaniacs (there is a proper sphere for science, and it is not to be defined by scientists). But it is in forgetting the stories, I want to suggest in closing, that the real work gets done.

In a 1999 essay entitled "Are Whales Fish?" the distinguished philosopher of science John Dupré took up a long-running philosophical argument concerning the relationship between ordinary language classification of living creatures and their formal scientific arrangement.[30] Should we expect "folkbiological" categories to converge with scientifically recognized kinds?[31] Are there good reasons for this to occur, or do instances of this phenomenon in the history of language-use reflect the unwarranted capitulation of perfectly real and operational taxonomies to inappropriately privileged classifications that have been elevated on the rising tide of a general "scientism"?[32] For Dupré, who advocates a "pluralistic view of biological classification" as part of a more general commitment to a "promiscuous realism" (which he situates in the middle of the slippery slope between essentialist and constructivist stances), there is no reason to expect folk classifications to be merely "first approximations" of taxonomic groups that can be more clearly delineated by scientists.[33] As he puts it, "[s]cientific classifications . . . are driven by specific, if often purely epistemic, purposes, and there is nothing fundamentally distinguishing such purposes from the more mundane rationales underlying folk classifications."[34] Thus it is Dupré's ongoing project to "claim that

---

[30] John Dupré, "Are Whales Fish?" in *Folkbiology,* edited by Scott Atran and Douglas L. Medin (Cambridge, MA: MIT Press, 1999), pp. 461–476. For the larger philosophical argument that informs this short piece, see John Dupré, *The Disorder of Things: Metaphysical Foundations of the Disunity of Science* (Cambridge, MA: Harvard University Press, 1993). Also relevant is idem, "Natural Kinds and Biological Taxa," *Philosophical Review* 90, no. 1 (1981), pp. 66–90. My brief discussion below of philosophical approaches to these problems is no substitute for reading the literature with which Dupré is engaged. See, for instance: Carl Gustav Hempel, "Fundamentals of Taxonomy," chapter 6 of his *Aspects of Scientific Explanation, and Other Essays in the Philosophy of Science* (New York: Free Press, 1965); Philip Kitcher, "Species," *Philosophy of Science* 51, no. 2 (1984), pp. 308–333; and Elliott Sorber, "Sets, Species, and Evolution: Comments on Philip Kitcher's 'Species,'" *Philosophy of Science* 51, no. 2 (1984), pp. 334–341. A philosophically sensitive historian's take on some of these issues can be found in Gordon McOuat, "From Cutting Nature at Its Joints to Measuring It: New Kinds and New Kinds of People in Biology," *Studies in the History and Philosophy of Science* 32, no. 4 (2001), pp. 613–645.

[31] For a detailed empirical investigation of this question, see Brent Berlin, *Ethnobiological Classification: Principles of Categorization of Plants and Animals in Traditional Societies* (Princeton: Princeton University Press, 1992).

[32] Dupré himself invokes a misguided scientism: Dupré, "Are Whales Fish?" p. 472.

[33] Ibid., pp. 461 and 472.

[34] Ibid., p. 462.

folk classifications should be treated on a par with scientific classifications," since there are consistently "good and obvious reasons for our
ordinary language distinctions," and no clear reason always and immediately to permit narrowly tailored epistemic or genealogical preoccupations
to subvert the implicit intentions of ordinary-language categories.[35]

As his title suggests, Dupré investigates this problem by taking up
the question of whether whales, properly speaking, "are" fish. And while
he must acknowledge that educated usage among English speakers on
the cusp of the third millennium obliges him to concede that they are
not, their displacement from the category rankles him, since he can
demonstrate that they once were, and he can see no compelling reason
why they should have lost that status. After all, neither "whale" nor "fish"
is *really* a proper scientific group: under scrutiny, both turn out to be
"biologically arbitrary," since "whale" is generally taken to mean only the
large cetaceans, and must therefore trace an amoeba-shaped region on
the textbook evolutionary tree of cetaceans (to grab the large odontocetes like the sperm whale and join them with the mysticetes, while ex-
cluding all the smaller dolphins and porpoises), and "fish" is probably the
most familiar example of what biologists now call a "paraphyletic" group,
since from the perspective of cladistics, the "fish" branch of creation includes all terrestrial vertebrates (which are, in an evolutionary sense, air-
breathing lobe-finned fish that took to living on land not very long
ago).[36] By these lights we're all fish. This puts Dupré in a position to
note: "Given, then, that neither 'whale' nor 'fish' is really a scientific term,
the rationale for the dictum taught religiously to all our children that
whales are not fish . . . is more than a little unclear."[37] He can then extend
this observation into an exemplar of the larger philosophical problem he
is after: "Indeed, the argument that because whales are mammals they
cannot be fish seems to me to be a paradigm for the confusion between
scientific and ordinary language biological kinds."[38] Why not mammalian fish?

All of which leaves him with a properly historical question—How
did whales become non-fish? which looms over the normative one
that he sets out to answer (Ought they have done so?). As a philosopher,

[35] Ibid., pp. 462 and 464.
[36] Omitting the lobe-finned fish leaves the category of "ray-finned fish," which is a mono-
phyletic group.
[37] Dupré, "Are Whales Fish?" p. 466.
[38] Ibid., p. 468.

Dupré has a taste for conjectural history, and thus he puts aside an actual "historical study of the usage" and allows himself an historical hypothesis: "For want of a better explanation, I conclude that the exclusion of whales from the category of fish developed as a response to greater scientific knowledge of the nature of whales, which was, as I have argued, a somewhat misguided response."

I would like to offer this book as an extended argument for the proposition that there is no substitute for actually *doing* the history. For while one might indeed speculate that whales ceased to be fish because of some general post-Enlightenment fawning over scientific expertise (however relevant or justified), a close look at how whales actually became non-fish in New York in the early nineteenth century has told a very different story: in fact what happened then and there was that scientific expertise took a terribly public bloody nose, and whales ceased to count as fish because of the behind-the-scenes legislative lobbying by a clique of oil merchants and chandlers, who managed to tweak the relevant statute right out from under public opinion, judicial scrutiny, and a jury verdict. By the time it was over "science" had been sent to the wings by all concerned.

The larger point is this: there is little reason to think we can reach reliable answers to the normative questions philosophers rightly like to ask and answer, without real and careful attention to the historical questions on which those answers often subtly hinge. For instance, "scientism" may have its explanatory functions, but we would do well to resist the ready temptation to call on it to do too much historical work, since it is a blunt instrument with which to investigate the complex relationship between science and society.

But perhaps this point is merely an instance of the kind of reciprocal and iterative rug-pulling that too often marks the exchanges between historians and philosophers. Is it possible to hope for something more? Maybe. As Simon Schaffer and others have recently argued, we do not simply need histories of philosophical problems and their solutions (nor do we want a blanket insistence that such problems would somehow be better solved historically), but rather we must seek philosophically informed histories of the examples, instances, and objects that philosophers have settled on, nurtured, and mobilized in making their arguments—histories, in other words, that refuse to let the history and the philosophy, the past and the present, the concrete and the abstract, fall neatly into their fixed cubbies in the taxonomy of the world and the intellect, but that instead show that taxonomy aborning. So Schaffer asks, for instance, how did "soap bubbles" meander back and forth between the domains of *vanitas* metaphysics and fluid mechanics, becoming the iconic objects of late-Victorian popular science, and later serving to exemplify certain privileged forms of evidence and rationality?[39] Histories like these, which insist that the instruments of abstraction are concrete, and that concrete instruments are simultaneously abstract, knit new connections between historical and philosophical investigation.[40]

In the case of the whale-fish it is possible to recover the elements of such a philosophical history, since *Maurice v. Judd* in fact lived a secret life in the nineteenth century as a philosophical *exemplum*, as a privileged problem for the consideration of scientific language and its place in social life. Though Dupré is not aware of it, his investigation of the relationship between science and society by means of the history of whales and fish itself has a history.

It would appear that the polymathic William Whewell—author of the massive *History of the Inductive Sciences*, custodian of the legacies of Bacon and Newton at Trinity College, Cambridge, and himself a common ancestor of historians and philosophers of science—came across the description of the case of *Maurice v. Judd* in his (English) edition of the *Règne Animal* when he was at work on the *Philosophy of the Inductive Sciences* (1840), the two-volume "lesson" to be gleaned from his ranging and exhaustive *History*.[41] In volume 1 of that philosophical treatise, enumerating (in Baconian style) "Aphorisms Concerning the Language of Science," Whewell revisited the fact pattern of *Maurice v. Judd* in consideration of the principle that "*Terms must be constructed and appropriated so as to be fitted to enunciate simply and clearly true general propositions*" (Aphorism VIII):

> The Nomenclature which answers the purposes of Natural History is a systematic nomenclature. . . . But we may remark that

[39] Simon Schaffer, "The Urbanity of Classical Physics: Soap Bubbles, Time Machines and Other Metropolitan Marvels," Lawrence Stone Visiting Professorship Lecture, Princeton University, 9 December 2003. A shortened version of this paper appears as Simon Schaffer, "A Science Whose Business Is Bursting: Soap Bubbles as Commodities in Classical Physics," in *Things That Talk: Object Lessons from Art and Science,* edited by Lorraine Daston (Cambridge, MA: Zone, 2004). This volume, which takes up a series of what might be called "philosophical objects," offers several examples of this sort of investigation.

[40] For a brief discussion of this approach, see Peter L. Galison and D. Graham Burnett, "Einstein, Poincaré, and Modernity," *Daedalus* 132, no. 2 (2003), pp. 41–55. For a thoughtful treatment of these problems from a philosopher's perspective: Ian Hacking, *Historical Ontology* (Cambridge, MA: Harvard University Press, 2002), particularly chapter 1.

[41] On Whewell generally, see Richard Yeo, *Defining Science: William Whewell, Natural Knowledge, and Public Debate in Early Victorian Britain* (Cambridge: Cambridge University Press, 1993).

the Aphorism now before us governs the use of words, not in science only, but in common language also. Are we to apply the name of *fish* to animals of the whale kind? The answer is determined by our present rule: we are to do so, or not, accordingly as we can best express true propositions.

This analysis led Whewell to offer what amounted to a philosophical ratification of the verdict in *Maurice v. Judd,* while rejecting any suggestion that ordinary language ought to submit to the razor of comparative anatomy:

> If we are speaking of the internal structure and physiology of the animal, we must not call them *fish;* for in these respects they deviate widely from the fishes: they have warm blood, and produce and suckle their young as land quadrupeds do. But this would not prevent our speaking of the *whale-fishery,* and calling such animals fish on all occasion connected with this employment; for the relations thus arising depend upon the animal's living in the water, and being caught in a manner similar to other fishes. *A plea that human laws which mention fish do not apply to whales, would be rejected at once by an intelligent judge.*[42]

Later in the same volume, characterizing the "idea of likeness" that governs the proper use of language, and that serves for him as the organizing principle of the classificatory sciences (here rejecting the notion that observed "kinds" are really only properly thought of as the product of exhaustive "definitions," and instead working toward the idea that a particular "type" gives shape to a group by means of "likeness"), Whewell again returns to the whale-fish problem, using it again to defend the plural domains within which the concept of likeness may work, even when a given term is thereby forced to lead multiple lives:

> The terms with which we are here most concerned are names of classes of natural objects; and when we say that the principle and the limit of such names are their use in expressing propositions concerning the classes, it is clear that much will depend on

---

[42]William Whewell, *The Philosophy of the Inductive Sciences, Founded upon Their History* (London: Parker, 1840), vol. 1, p. lxxv; last emphasis mine. I base my conjecture concerning the origin of Whewell's interest in this issue on the fact that the discussion of Aphorism VIII begins with a citation of Cuvier, and a precise reference to the *Règne Animal,* indicating that Whewell was reading in that volume as he composed this section, and making it seem quite likely that he took up the whale/fish question in response to the note on *Maurice v. Judd* in Griffith's edition of that work.

the kind of propositions which we mainly have to express: and that the same name may have different limits, according to the purpose we have in view. For example, is the whale properly included in the general term *fish?*

And again the answer is made to depend on context:

> When men are concerned in catching marine animals, the main features of the process are the same however the animals may differ; hence whales are classed with fishes, and we speak of the *whale-fishery.* But if we look at the analogies of organization, we find that, according to these, the whale is clearly not a fish, but a *beast,* (confining this term, for the sake of distinctness, to suckling beasts or *mammals*). In Natural History, therefore, the whale is not included among fish.[43]

For Whewell, then, the difference between the whale-fish and the non-fish-whale was the difference between the community of seamen and the community of naturalists, and (like Dupré, albeit for very different reasons) he displayed no particular taste for a convergence of usage, since the word "fish" was being used in each arena in such a way as to express true propositions, and thus the principle at stake—"that the condition of the use of terms is the possibility of general, intelligible, consistent assertions"—was fully satisfied at sea, in the courts, and at the *Muséum.*[44]

Whewell's effort to exemplify the proper operation of name-giving and category formation by means of a discussion of how to parse whales,

---

[43] This and the above quote from Whewell, *The Philosophy of the Inductive Sciences,* vol. 1, p. 456.

[44] Though only, to be fair, if you did not know much about fishing and whaling, since in point of fact "the main features of the process" of catching these creatures looked nothing alike. For an interesting analysis of the broader set of commitments that informed Whewell's ideas about scientific nomenclature, see Simon Schaffer, "The History and Geography of the Intellectual World: Whewell's Politics of Language," in *William Whewell: A Composite Portrait,* edited by Menachem Fisch and Simon Schaffer (Oxford: Clarendon Press, 1991), pp. 201–231. Also useful: Yeo, *Defining Science,* pp. 106–115. It is worth remembering how important these issues were in the 1830s and 1840s: just a few years later, in 1842, Hugh E. Strickland would deliver, through the British Association for the Advancement of Science, a set of formal *Rules of Zoological Nomenclature.* For an insightful account of the struggles around this effort (and the implications for ideas about species in the period), see Gordon McOuat, "Cataloguing Power: Delineating 'Competent Naturalists' and the Meaning of Species in the British Museum," *British Journal for the History of Science* 34 (2001), pp. 1–28. For the later history of these issues: Richard V. Melville, *Towards Stability in the Naming of Animals: A History of the International Commission on Zoological Nomenclature, 1985–1995* (London: International Trust for Zoological Nomenclature, 1995).

fish, and men immediately fell under the scrutiny of one of his closest readers, John Stuart Mill, who framed his *System of Logic* of 1843 as an extensive (if admiring) critique of Whewell's ideas about induction and the methods of scientific reasoning.[45] In discussing "classification, as subsidiary to induction," Mill himself took up the question of how different kinds of people might preoccupy themselves with different "kinds":

> [T]he classification of objects should follow those of their properties which indicate not only the most numerous, but also the most important peculiarities. What is here meant by importance? It has reference to the particular end in view: and the same objects, therefore, may admit with propriety of several different classifications. Each science or art forms its classifications of things according to the properties which fall within its special cognizance, or of which it must take account in order to accomplish its peculiar practical ends.

An acknowledgment that led to what might appear to be a version of the "pluralistic view of biological classification":

> A farmer does not divide plants, like a botanist, into dicotyledonous and monocotyledonous, but into useful plants and weeds. A geologist divides fossils, not, like a zoologist, into families corresponding to those of living species, but into fossils of the secondary and of the tertiary periods, above the coal and below the coal, & c. Whales are or are not fish, according to the purpose for which we are considering them.[46]

---

[45] At issue was a sharp polemic over the role of induction in the making and assessing of scientific hypotheses. Whether the "Mill-Whewell debate" can best be summarized as a showdown between deductivist (Whewell) and inductivist (Mill) approaches, or as a contest between rival conceptions of induction, remains unresolved among philosophers of science. The proper place of history and logic in the shaping of scientific theories was also clearly at issue. For an introduction to these problems, see: Gerd Buchdahl, "Deductivist versus Inductivist Approaches in the Philosophy of Science as Illustrated by Some Controversies between Whewell and Mill," in Fisch and Schaffer, *William Whewell*, pp. 312–344; Laura J. Snyder, "The Mill-Whewell Debate: Much Ado about Induction," *Perspectives on Science* 5, no. 2 (1997), pp. 159–198; and John Losee, "Whewell and Mill on the Relation between Philosophy of Science and History of Science," *Studies in the History and Philosophy of Science* 14, no. 2 (1983), pp. 113–126. Also useful (particularly for understanding the continuing relevance of the debate in current philosophy of science) is: Peter Achinstein, "Inference to the Best Explanation: Or, Who Won the Mill-Whewell Debate?" *Studies in the History and Philosophy of Science* 23, no. 2 (1992), pp. 349–364; this is an essay review of Peter Lipton's *Inference to the Best Explanation* (London: Routledge, 1991).

[46] John Stuart Mill, *A System of Logic, Ratiocinative and Inductive* (New York: Harper and Brothers, 1846), p. 435.

Mill even went on to quote in full Whewell's paragraph on this last issue, concluding with the observation that a judge might well deploy the fisherman's signification of "fish" when interpreting a law about whaling.[47]

But Mill appears to have been less certain about the long-term virtues of such pluralism and the "promiscuity" it encouraged. Turning from Whewell's example of whale-mammals in comparative anatomy and whale-fish on the high seas, Mill recalled the distinctiveness and privileged significance of approaching whales or other objects "not for any special practical end, but for the sake of extending our knowledge of the whole of their properties and relations." In this special exercise what came to the fore were those "most important attributes" which "contribute most, either by themselves or by their effects, to render the things like one another, and unlike other things." By paying attention to those characteristics of objects "which would most impress the attention of a spectator who *knew all their properties but was not specially interested in any*," the true philosopher of nature could discover the attributes "which fill, as it were, the largest space in their existence," and thus mark out the classes properly called "natural groups."[48] Hence he who approached a whale with *no particular end in mind* stood the best chance of grasping the nature of whales, and thus of offering a truly scientific classification of organic nature based on the true principles of rational classification.

---

[47] Both Mill and Whewell were generally interested in the relationship between legal and scientific forms of evidence, and both discuss the ways that judges and advocates reason about decisions. See Katharine Anderson, *Predicting the Weather: Victorians and the Science of Meteorology* (Chicago: University of Chicago Press, 2004), particularly chapter 1. It is also interesting to note that between the publication of Whewell's *Philosophy of the Inductive Sciences* and Mill's *System of Logic*, the "Error of Supposing the Whale to Be a Fish" had run in block caps as a title in the *Times of London* (1 October 1841, p. 7). The occasion was a review of *Parley's Penny Library* (London: Cleave, 1841), the first volume of a series of didactic Victorian juvenile novellas by the pseudonymous "Peter Parley," which devoted an entire chapter (chapter 13, complete with anatomical illustrations) to the domestic scene of a dutiful father disabusing his children (and wife) of their confusion about the taxonomic status of the cetaceans. It is striking to consider that, in the end, the origin of the "dictum taught religiously to all our children that whales are not fish" (Dupré, "Are Whales Fish?" p. 466) may well have as much to do with the rise of Victorian hackwork journalism as with the rise of comparative anatomy (after all, one of the leading nineteenth-century spokesmen for systematic taxonomy, Whewell, would have been perfectly happy with Melville's definition of the whale: "a spouting fish with a horizontal tail"). This would confirm the importance of seeing "popular" and "professional" science in the Victorian period as reciprocally constituted to a surprising degree. See: Bernard Lightman, ed., *Victorian Science in Context* (Chicago: University of Chicago Press, 1997); and James Secord, *Victorian Sensation: The Extraordinary Publication, Reception, and Secret Authorship of* Vestiges of the Natural History of Creation (Chicago: University of Chicago Press, 2000).

[48] This and the above quotes from: Mill, *A System of Logic*, p. 436; emphasis mine.

In the process it would become possible to make good inductions about the membership of classes so constructed.[49]

Moreover, for Mill, the virtues of this indifferently-interested view and the system it engendered recommended broader application. The principles on evidence in natural history taxonomy, he explained, "are applicable to all cases in which mankind are called upon to bring the various parts of any extensive subject into mental coördination," and therefore they "are as much to the point when objects are to be classed for the purposes of art or business, as for those of science." Indeed, "[t]he proper arrangement, for example, of a code of laws, depends upon the same scientific conditions as the classifications in natural history; nor could there be a better preparatory discipline for that important function, than the study of the principles of a natural arrangement, not only in the abstract, but in their actual application to the class of phenomena for which they were first elaborated."[50]

Here was an argument that would make Mitchill the tutor in the Mayor's Court, and thereby entail the resolution of the social conflict of *Maurice v. Judd* not in the Swamp, or in Albany, or anywhere on the Atlantic shores or the waters that washed them, but rather on the chessboard of natural historical classification, an even surface mysteriously raised above the world of mere people and places. It was the argument soundly rejected in New York in 1818. But in the end, it may be the argument that has won.

Traced into the filiations of philosophical citation, the afterlife of *Maurice v. Judd* is revealed to have been the occasion of reflection not merely

---

[49] It is here that this "other" Mill-Whewell debate (about the proper basis for taxonomy, and the propriety of pluralism in nomenclature) intersects with the Mill-Whewell debate more familiar to the philosophers. Induction and deduction alike require meaningful sets upon which to operate, and the making of meaningful sets is the business of classification. Where classification is concerned, Mill and Whewell ultimately divide, it seems to me, on the question of whether a "Natural group" is best given by a "type" (Whewell) or by a "definition" (Mill). A slightly different terminology might contrast "family resemblance" categories (though it could be argued that Whewell thinks of his types more like Platonic forms) with those derived from an enumeration of characteristics. The latter approach, which Mill borrows from Comte, inclines toward algorithmic processes, the former toward something like connoisseurship. The contrast is, I think, instructive in considering the larger implications of Mill's and Whewell's ideas about the proper relationship between science and society, but I must leave it to the real philosophers of science to play out whether more attention to this issue would shift accounts of the Mill-Whewell debate over induction itself.

[50] Mill, *A System of Logic*, p. 447.

on whales and fish, but on the proper relationship between zoology and law, science and society.[51] Could a judge rightly consider a whale a fish for the purpose of statutory interpretation? Perhaps, but that same judge would do well to study the *Règne Animal* (strangely) in order to learn the proper basis for his judgment concerning categories, and thereby to practice the exercise of right reason. Need the folk taxonomies of labor and the market converge with the classifications of natural history? Perhaps not, though the essence of a natural class would only ever be revealed by the class of citizens who came to nature with no end in view, and whose view was thereby accorded the final word.

On display in this debate, then, is the operation of the very move that has deeply preoccupied historians of science in the last several decades: the strange phenomenon by which the crisp caesura between science and society is articulated silently, by an ellipsis that stands in the place of forgotten histories. Thus we see an "asocial" knowledge (a knowledge that has no hunger, needs no salary) commandingly define itself in a perpetual sequence of contrasts to socially blinkered views (of whalemen, tanners, chandlers), and in doing so become a socially *privileged* form of knowledge: austerely asocial as it enters the domain of social conflict, science is promiscuously social in the constitution of its asocial identity. In the process, "science" becomes something to be contrasted with "society," and becomes at the same time a resource upon which society can rely to solve social problems.

In reflecting on the relationship between history and philosophy, we might hazard a parallel construction, since philosophy can plausibly be seen as austerely ahistorical as it enters the domain of historical debate, but promiscuously historical in the constitution of its ahistorical identity. Thus Dupré attacks the historical question "are whales fish?" careful not to touch the past, even as he is unfolding a philosophical example

---

[51] Nor did things come to an end with Whewell and Mill: Thomas Henry Huxley, writing at the end of the nineteenth century, returned to the whale-fish problem when he needed to explain what "philosophical" anatomy was, and what it meant to grasp the "totality" of "resemblances" (see Huxley, "Owen's Position in the History of Anatomical Science"). William Henry Flower, Keeper of Zoology at the British Museum, not only wrote about why whales were not fish, but also played out the legal implications of this fact in a lecture to the Royal Colonial Institute in 1895. And, as proof that history unfolds according to its own mysterious cycles, in 1919, almost a century to the day after *Maurice v. Judd*, the American zoologist Frederic A. Lucas, director of the American Museum of Natural History, was called to give testimony before the Board of Appraisers in New York City to "show why a whale was not a fish"—it appears that an American whaling company was trying to avoid paying a duty on whale meat as a "fish product." See Lucas to Harmer, 10 March 1919, Scott Polar Research Institute Archives, MS 1284/2, Falkland Islands Committee, Correspondence.

that has been made philosophical by and through (an elided) history. In the process "philosophy" comes to be distinguished from "history" by a process of erasure not entirely unlike the process by which "science" comes to be distinguished from "society." Along the way, philosophers and scientists (and indeed, complicitous historians) win themselves sets of tools with which to work upon both the past and the present.

Is a whale a fish? Is science social? Is philosophy historical? The precedent question is always this: What stories must be forgotten to answer these questions?

Acknowledgments

This book benefited from the generous attentions of friends, colleagues, students, and family. Princeton University supported my research with leave and travel grants; a National Endowment for the Humanities Fellowship in 2003–2004 (project number 205-6058) permitted me to draft the manuscript, which was completed during my tenure of the Christian Gauss Preceptorship. A number of institutions facilitated the research represented here, including the New-York Historical Society, the New York Academy of Sciences, the New York Public Library, the American Museum of Natural History, the American Philosophical Society, the Historical Society of Pennsylvania, the New York State Archives (Albany), the Old Records Division of the New York State Supreme Court, the Pennypacker Collection of the East Hampton Public Library, the Research Library and Kendall Institute of the New Bedford Whaling Museum, the G. W. Blunt White Library of Mystic Seaport, the Archives of the Library of Congress, and the Special Collections of Columbia University Libraries. Without the extraordinary services of the staff of Article Express and Interlibrary Services at Princeton, this work would have been impossible. The following individuals directly contributed to my work on this project, and I greatly appreciate the time and advice they gave: Bruce Adams, Danielle Allen, Katharine Anderson, Whitney Bagnell, Fred Bassett, Whitfield J. Bell, Thomas Bender, Elizabeth Bennett, Rachel Berg, Dan Bouk, Robert Bruesewitz, Joel Burlingham, Christina D. Burnett, Joyce Chaplin, Robert Cox, Angela Creager, Jeff Dolven, Ermenio D'Onofrio, Ann Downer-Hazell, Ariela Dubler, Michael Dyer, Richard Ellis, Paul Farber, Gretchen Feltes, Jim Folts, Stuart Frank, Anthony Grafton, Charles Greifenstein, Hendrik Hartog, Abby Heald, Alice Hudson, Jonathan Katz, Joshua Katz, Dorothy King, Bernard Lightman, John Logan, Michael Mahoney, Ed Mitchell, Fred Moreno, Joe Nardello, Lynn Nyhardt, Hershel Parker, Philip Pauly, Randy Reeves, Harriet Ritvo, Simon Schaffer, James Schulz, Jeff Schwegman, Jim Secord, Tim Smith, Alistair Sponsel, Christine Stansell, Helen Tilley, Mariam Touba, Marci Vail, Joseph Van Nostrand, John Walden, Sean Wilentz, and John Witt. In addition, helpful audiences at the Huntington Library's "Beasts of Sea and Land" conference, the 29th Whaling History Symposium at the New Bedford Whaling Museum, the Johns Hopkins History of Science Colloquium, the Columbia Law School Legal History Workshop, the Committee on Social Thought at the University of Chicago, the Center for Advanced Studies at the University of Illinois, the University of Toronto, and Princeton's History of Science Program Seminar provided invaluable criticism and commentary. Students in two graduate seminars, "Science Across the Seas" and "Humans and Animals," sharpened my thinking on the secondary literature. The committed editors at Princeton University Press helpfully brought the book to its present form. To all of these people and places I extend my sincere thanks. My wife, Christina, is my favorite legal historian and my first inspiration in all things.

Bibliography

First footnote citations are complete; subsequent references use author and short title. Because authors and titles were used irregularly in period newspapers, those references are always given in their entirety in the notes, and are not included in this bibliography. I made use of original newspaper collections at the New-York Historical Society, Princeton University, and the Library of Congress, in addition to consulting the extensive digital collection "America's Historical Newspapers, 1690–1922" (whose standard titles I have adopted). The abbreviation *IWF* stands for William Sampson, *Is a Whale a Fish?* (New York: Van Winkle, 1819).

UNPUBLISHED PRIMARY SOURCES (BY ARCHIVE)

Collection information included where relevant; specific citations in the footnotes.

American Philosophical Society Manuscript Library
    Wyck Collection
        Papers of Reuben Haines III
Columbia University, Rare Books and Manuscript Library
    DeWitt Clinton Papers
    Columbiana Collection
East Hampton Library
    Long Island Collection
        Mitchell [*sic*], Samuel Latham, Ms. Lectures on Zoology
Historical Society of Pennsylvania
    Carey Papers
    Gratz Collection
Library of Congress, Manuscript Division
    Peter Force Collection
        Jeremy Robinson Papers
    James Kent Papers
    Catherine Mitchill Papers
    William Sampson Family Papers
New Bedford Whaling Museum, Research Library
    Kendall Collection
    Old Dartmouth Historical Society Collection
New York Academy of Sciences
    Minutes of the Lyceum of Natural History
New-York Historical Society
    Joseph Bulkey Docket Book
    Cases of Mayor's Court 1797–1825
    DeWitt Clinton Daybook

Issacher Cozzens Portfolio Print Collection
Rufus King Papers
Meigs Papers
Minute Book of the New-York Historical Society
Minute Book of the New-York Literary and Philosophical Society
Pintard Papers
New York Public Library
  Manuscript Division
    Thomas A. Brayton Notebook
  Map Division
    Irma and Paul Milstein Division
  Print Collection
New York State Archives (Albany)
  Gideon Lee Papers
New York Supreme Court, County Clerk Archives
  Old Records Room
    Mayor's Court Docket, etc.
Scott Polar Research Institute Archives (Cambridge, UK)
  Falkland Islands Committee, Correspondence
Yale University, Manuscripts and Archives, Sterling Memorial Library
  Burr Family Papers
  Natural Science Manuscripts Collection
    John Torrey Papers
  Silliman Family Papers

PUBLISHED SOURCES

Exclusive of newspaper articles, for which full citations will be found in the notes.

A Correspondent. "Cultivation of Natural History in the University College of New York." *American Medical and Philosophical Register* 2 (1811): 154–163.

Aberbach, Alan David. *In Search of an American Identity: Samuel Latham Mitchill, Jeffersonian Nationalist.* New York: Peter Lang, 1988.

Accum, Friedrich Christian. *A Treatise on Adulterations of Food and Culinary Poisons.* London: J. Mallett, 1820.

Achinstein, Peter. "Inference to the Best Explanation: Or, Who Won the Mill-Whewell Debate?" *Studies in History and Philosophy of Science* 23, no. 2 (1992): 349–364.

Albion, Robert. *The Rise of New York Port, 1815–1860.* New York: Scribner's Sons, 1939.

Alderson, William T., ed. *Mermaids, Mummies, and Mastodons: The Emergence of the American Museum.* Washington, DC: Association of American Museums, 1992.

Aldrovandi, Ulisse. *De Piscibus.* Bononiae: Apud Bellagambam, 1613.

Allen, David Elliston. "Natural History and Social History." *Journal of the Society for the Bibliography of Natural History* 7, no. 4 (1976): 509–516.

———. *The Naturalist in Britain: A Social History.* Princeton: Princeton University Press, 1976.

Alschuler, Albert W. "Rediscovering Blackstone." *University of Pennsylvania Law Review* 145, no. 1 (1996): 1–55.

Anderson, Katharine. *Predicting the Weather: Victorians and the Science of Meteorology.* Chicago: University of Chicago Press, 2004.

Appel, Toby. *The Cuvier-Geoffroy Debate: French Biology in the Decades before Darwin.* New York: Oxford University Press, 1987.

———. "Science, Popular Culture, and Profit: Peale's Philadelphia Museum." *Journal of the Society for the Bibliography of Natural History* 9, no. 4 (1980): 619–634.

Ashton, Jean. "Tall Tales and Whales: Wonders of Barnum's Museum." *Imprint* 16, no. 2 (Autumn 1991): 15–25.

Atran, Scott. *Cognitive Foundations of Natural History: Towards an Anthropology of Science.* Cambridge: Cambridge University Press, 1990.

Atran, Scott, and Douglas L. Medin, eds. *Folkbiology.* Cambridge, MA: MIT Press, 1999.

Baatz, Simon. "Knowledge, Culture, and Science in the Metropolis: The New York Academy of Sciences, 1817–1970." *Annals of the New York Academy of Sciences* 584 (1990): 1–269.

———. "Philadelphia Patronage: The Institutional Structure of Natural History in the New Republic, 1800–1833." *Journal of the Early American Republic* 8, no. 2 (1988): 111–138.

Barker, Veronica F., and Ian A. D. Bouchier. "The Polymath Practitioners." *The Practitioner* 217 (1976): 428–434.

Barnum, Phineas T. *The Life of P. T. Barnum.* New York: Redfield, 1855.

Barthelmess, Klaus, and Joachim Münzig, *Monstrum Horrendum.* 3 vols. Bremerhaven: Deutsches Schiffahrtsmuseum, 1991.

Barton, Benjamin Smith. *Discourse on Some of the Principal Desiderata in Natural History and on the Best Means of Promoting the Study of This Science, in the United States.* Philadelphia: Denham and Town, 1807.

Bates, Ralph S. *Scientific Societies in the United States.* Cambridge, MA: MIT Press, 1965.

Beale, Charles Currier. *William Sampson, Lawyer and Stenographer.* Little Visits to the Homes of Eminent Stenographers, no. 1. Boston: n.p., 1907.

Beale, Thomas. *A Few Observations on the Natural History of the Sperm Whale.* London: Effingham Wilson, 1835.

———. *The Natural History of the Sperm Whale.* London: J. Van Voorst, 1839.

Bender, Thomas. *New York Intellect: A History of Intellectual Life in New York City from 1750 to the Beginnings of Our Own Time.* New York: Knopf, 1987.

———, ed. *The University and the City: From Medieval Origins to the Present.* New York: Oxford University Press, 1988.

Benes, Peter. "To the Curious: Bird and Animal Exhibitions in New England, 1716–1825." In *New England's Creatures: 1400–1900,* edited by Peter Benes. Boston: Boston University, 1995.

———, ed. *New England's Creatures: 1400–1900.* Boston: Boston University, 1995.

Bennett, Frederick Debell. *Narrative of a Whaling Voyage round the Globe from the Year 1833 to 1836.* 2 vols. London: Richard Bentley, 1840.

Berlin, Brent. *Ethnobiological Classification: Principles of Categorization of Plants and Animals in Traditional Societies.* Princeton: Princeton University Press, 1992.

———. "Folk Systematics in Relation to Biological Classification and Nomenclature." *Annual Review of Ecology and Systematics* 4 (1973): 259–271.

Bicheno, J. E. "On Systems and Methods in Natural History." *Transactions of the Linnaean Society of London* 15 (1827): 479–496.

Bigland, John. *Letters on Natural History.* London: James Cundee, 1810.

Bloomfield, Maxwell. *American Lawyers in a Changing Society, 1776–1876.* Cambridge, MA: Harvard University Press, 1976.

Bloor, David. "Durkheim and Mauss Revisited: Classification and the Sociology of Knowledge." *Studies in History and Philosophy of Science* 13, no. 4 (1982): 267–297.

Blum, Ann Shelby. *Picturing Nature: American Nineteenth-Century Zoological Illustration.* Princeton: Princeton University Press, 1993.

Bonnaterre, Pierre Joseph. *Cétologie.* Paris: Chez Panckoucke, 1789.

Bornbusch, Alan H. "Lacépède and Cuvier: A Comparative Case Study of Goals and Methods in Late Eighteenth- and Early Nineteenth-Century Fish Classification." *Journal of the History of Biology* 22, no. 1 (Spring 1989); 141–161.

Bouchier, Ian A.D. "Whales and Whaling: Contributions by the Medical Profession." *Medical History* 27, no. 2 (April 1983): 182–189.

Bouk, Dan, and D. Graham Burnett. "Knowledge of Leviathan: Charles W. Morgan Anatomizes His Whale." Forthcoming in the *Journal of the Early Republic.*

Brewster, David, ed. *New Edinburgh Encyclopedia.* "Second American Edition." New York: Whiting and Watson, 1814.

Brigham, David R. "Social Class and Participation in Peale's Philadelphia Museum." In *Mermaids, Mummies, and Mastodons: The Emergence of the American Museum,* edited by William T. Alderson. Washington, DC: American Association of Museums, 1992.

Brisson, Jacques. *Ornithologie.* 6 vols. Paris: Bauche, 1760.

———. *Le Règne Animal Divisé en IX Classes.* Paris: Bauche, 1756.

Brown, Chandos Michael. *Benjamin Silliman: A Life in the Young Republic.* Princeton: Princeton University Press, 1989.

———. "A Natural History of the Gloucester Sea Serpent: Knowledge, Power, and the Culture of Science in Antebellum America." *American Quarterly* 42, no. 3 (1990): 402–436.

Browne, Irving. "William Sampson." *The Green Bag* 3, no. 8 (1896): 313–325.

Browne, J. Ross. *Etchings of a Whaling Cruise.* Cambridge, MA: Harvard University Press, Belknap Press, 1968 [1846].

Browne, Sir Thomas. *Selected Writings.* Edited by Claire Preston. New York: Routledge, 2003.

Buchdahl, Gerd. "Deductivist versus Inductivist Approaches in the Philosophy of Science as Illustrated by Some Controversies between Whewell and Mill." In *William Whewell: A Composite Portrait,* edited by Menachem Fisch and Simon Schaffer. Oxford: Clarendon Press, 1991.

Burkhardt, Richard W. Jr. *The Spirit of System: Lamarck and Evolutionary Biology.* Cambridge, MA: Harvard University Press, 1977.

Burnett, D. Graham. "Hydrographic Discipline." In *The Imperial Map,* edited by James Akerman. Chicago: University of Chicago Press, forthcoming.

Burnett, D. Graham, et al. "Science and the Law." *Isis* 98, no. 2 (2007): 310–350.

Burrows, Edwin G., and Mike Wallace. *Gotham: A History of New York City to 1898.* New York: Oxford University Press, 1999.

Busch, Briton Cooper. "Elephants and Whales: New London and Desolation, 1840–1900." *American Neptune* 40, no. 2 (1980): 117–126.

———. *Whaling Will Never Do for Me: The American Whaleman in the Nineteenth Century.* Lexington: University Press of Kentucky, 1994.

Bynum, W. F., and Roy Porter. *William Hunter and the Eighteenth-Century Medical World.* Cambridge: Cambridge University Press, 1985.

Byrne, James. "Cetus and Balena." Unpublished paper, 2003.

Camper, Pierre. *Observations Anatomiques sur la Structure Intérieure et le Squelette de Plusieurs Espèces de Cétacés.* With notes by Georges Cuvier. Paris: Gabriel Dufour, 1820.

Cartmill, Matt. *A View to a Death in the Morning: Hunting and Nature through History.* Cambridge, MA: Harvard University Press, 1993.

Casarino, Cesare. *Modernity at Sea: Melville, Marx, Conrad in Crisis.* Minneapolis: University of Minnesota Press, 2002.

Chaplin, Joyce E. "Nature and Nation: Natural History in Context." In *Stuffing Birds, Pressing Plants, Shaping Knowledge,* edited by Sue Ann Prince. Philadelphia: American Philosophical Society, 2003.

Chase, Jackson H., and George W. Heaton [Jones, pseud.]. *Life and Adventure in the South Pacific. By a Roving Printer.* New York: Harper and Brothers, 1861.

Chase, Owen, et al. *Narratives of the Wreck of the Whale-Ship Essex.* New York: Dover, 1989.

Cheever, Henry T. *The Whale and His Captors.* New York: Harper and Brothers, 1849.

"Classification of Mammalia." *The Zoological Magazine, or Journal of Natural History* 1, no. 6 (1833): 176–190.

Clinton, DeWitt. "An Introductory Discourse." *Transactions of the Literary and Philosophical Society of New-York* 1 (1815): 19–184.

——— [Hibernicus, pseud.]. *Letters on the Natural History and Internal Resources of the State of New-York.* New York: E. Bliss and E. White, 1822.

Cloud, Enoch Carter. *Enoch's Voyage: Life on a Whaleship, 1851–1854.* Edited by Elizabeth McLean. Wakefield, RI: Moyer Bell, 1994.

Coenen, Adriaen. *The Whale Book: Whales and Other Marine Animals as Described by Adriaen Coenen in 1585.* Edited by Florike Egmond and Peter Mason. London: Reaktion, 2003.

Cohen, Claudine. *The Fate of the Mammoth: Fossils, Myth, and History.* Chicago: University of Chicago Press, 2002.

Cole, F. J. *A History of Comparative Anatomy from Aristotle to the Eighteenth Century.* London: MacMillan, 1944.

College of Physicians and Surgeons in the City of New York. *A Concise Memorandum of Certain Articles Contained in the Museum of Samuel L. Mitchill.* New York: E. Conrad, circa 1825.

Collison, Robert. *Encyclopaedias: Their History throughout the Ages.* New York: Hafner, 1964.

Colnett, James. *A Voyage to the South Atlantic and round Cape Horn into the Pacific Ocean.* New York: Da Capo Press, 1968 [1798].

Combist, Frum O. "Maycomb-MacDonnell, M. I. Return (1764–1818?)." In *The Dictionary of Nineteenth-Century British Scientists,* edited by Bernard Lightman. Chicago: University of Chicago Press, 2004.

Cooley, Thomas M. *A Treatise on the Constitutional Limitations Which Rest upon the Legislative Power of the States of the American Union.* Boston: Little and Brown, 1868.

Cooper, James Fenimore. *The Sea Lions.* Amsterdam: Fredonia, 2002 [1849].

Corey, James Robert. "Herman Melville and the Theory of Evolution." Ph.D. diss., Washington State University, 1968.

Cornog, Evan. *The Birth of Empire: DeWitt Clinton and the American Experience, 1769–1828.* New York: Oxford University Press, 1998.

Coulson, Thomas. *Joseph Henry: His Life and Work.* Princeton: Princeton University Press, 1950.

Creager, Angela N. H., and William Chester Jordan, eds. *The Animal/Human Boundary: Historical Perspectives.* Rochester, NY: University of Rochester Press, 2002.

Creighton, Margaret S. *Rites and Passages: The Experience of American Whaling, 1830–1870.* Cambridge: Cambridge University Press, 1995.

Croizat, Leon. "History and Nomenclature of the Higher Units of Classification." *Bulletin of the Torrey Botanical Club* 72, no. 1 (1945): 52–75.

Cuvier, Frédéric. *De l'Histoire Naturelle des Cétacés.* Paris: Librairie Encyclopédique de Roret, 1836.

Cuvier, Georges. *The Animal Kingdom: Arranged in Conformity with Its Organization, with Supplementary Additions to Each Order by Edward Griffith.* 15 vols. London: Whittaker and Co., 1827–1835.

———. *Essay on the Theory of the Earth.* Edited by Robert Jameson, with additions and a new introduction by Samuel Latham Mitchill. New York: Kirk and Mercein, 1818.

———. *Le Règne Animal.* 4 vols. Paris: Déterville, 1817.

———. *Le Règne Animal.* 2nd ed. 5 vols. Paris: Déterville, 1829.

Dain, Bruce R. *A Hideous Monster of the Mind: American Race Theory in the Early Republic.* Cambridge, MA: Harvard University Press, 2002.

Dannenfeldt, K. H. "Ambergris: The Search for Its Origin." *Isis* 73, no. 3 (1982): 382–397.

Daniels, George H., ed. *Nineteenth-Century American Science: A Reappraisal.* Evanston: Northwestern University Press, 1972.

———. "The Process of Professionalization in American Science: The Emergent Period, 1820–1860." *Isis* 58, no. 2 (1967): 150–166.

Darwin, Charles. *On the Origin of Species: A Facsimile of the First Edition.* Cambridge, MA: Harvard University Press, 1964 [1859].

Daston, Lorraine, ed. *Things That Talk: Object Lessons from Art and Science.* Cambridge, MA: Zone, 2004.

Daston, Lorraine, and Gregg Mitman, eds. *Thinking with Animals: New Perspectives on Anthropomorphism.* New York: Columbia University Press, 2004.

Daudin, Henri. *Cuvier et Lamarck: Les Classes Zoologiques et l'Idée de Série Animale 1790–1830.* Paris: F. Alcan, 1926.

Davis, David Brion. *Antebellum American Culture: An Interpretive Anthology.* Lexington, MA: Heath, 1979.

Davis, Lance E., Robert E. Gallman, and Karin Gleiter. *In Pursuit of Leviathan: Technology, Institutions, Productivity, and Profits in American Whaling, 1816–1906.* Chicago: University of Chicago Press, 1997.

Davis, William M. *Nimrod of the Sea; or, The American Whaleman.* London: Sampson Low, Marston, Low, and Searle, 1874.

de Wailly, Noël François. *Nouveau Vocabulaire François.* 2nd ed. Paris: Rémont, 1803.

Delaney, David. *Law and Nature.* Cambridge: Cambridge University Press, 2003.

Delbanco, Andrew. *Melville: His World and Work.* New York: Knopf, 2005.

Desmond, Adrian, and James R. Moore. *Darwin.* New York: Norton, 1994.

Dewhurst, Henry William. *The Natural History of the Order Cetacea, and the Oceanic Inhabitants of the Arctic Regions.* London: Published by the Author, 1834.

Dierig, Sven, Jens Lachmund, and Andrew Mendelsohn, eds. *Science and the City. Osiris,* 2nd ser., 18 (2003).

DiGregorio, Mario A. "In Search of the Natural System: Problems of Zoological Classification in Victorian Britain." *History and Philosophy of the Life Sciences* 4 (1982): 225–254.

"Donations for the Library Cabinet." *Transactions of the Literary and Philosophical Society of New-York* 1 (1815): backmatter.

Dow, George Francis. *Whale Ships and Whaling: A Pictorial History.* New York: Dover Publications, 1985.

Dowling, William C. *Literary Federalism in the Age of Jefferson.* Columbia: University of South Carolina Press, 1999.

Drake, Joseph Rodman. *The Croakers.* New York: The Bradford Club, 1860.

———. *Poems, by Croaker, Croaker and Co. and Croaker, Jun.* New York: For the Reader, 1819.

Dudley, Paul. "An Essay upon the Natural History of Whales." *Philosophical Transactions of the Royal Society* 33 (1725): 256–269.

Dupré, John. "Are Whales Fish?" In *Folkbiology,* edited by Scott Atran and Douglas L. Medin. Cambridge, MA: MIT Press, 1999.

———. *The Disorder of Things: Metaphysical Foundations of the Disunity of Science.* Cambridge, MA: Harvard University Press, 1993.

———. "Natural Kinds and Biological Taxa." *Philosophical Review* 90, no. 1 (1981): 66–90.

Durkheim, Emile, and Marcel Mauss. "De Quelques Formes Primitives de Classification." *Année Sociologique* 6 (1903): 1–71.

Dyer, Michael P. "The Historical Evolution of the Cutting-In Pattern, 1798–1876." *American Neptune* 59, no. 2 (1999): 134–149.

Elliot, Clark A. *History of Science in the United States: A Chronology and Research Guide.* New York: Garland, 1996.

Ellsworth, Lucius F. *Craft to National Industry in the Nineteenth Century: A Case Study of the Transformation of the New York State Tanning Industry.* New York: Arno Press, 1975.

Ely, Ben-Ezra Stiles. *"There She Blows": A Narrative of a Whaling Voyage.* Middletown, CT: Wesleyan University Press, 1971 [1849].

Ereshefsky, Marc. "The Evolution of the Linnaean Hierarchy." *Biology and Philosophy* 12 (1997): 493–519.

Eschricht, Daniel Fredrik. "Sur une Nouvelle Méthode de l'Etude des Cétacés." *Comptes Rendus des Séances de l'Académie des Sciences* 47 (1858): 51–60.

Evans, Martin H. "Statutory Requirements regarding Surgeons on British Whale-Ships." *Mariner's Mirror* 91, no. 1 (2005): 7–12.

Fairchild, Herman L. *History of the New York Academy of Sciences.* New York: The Author, 1887.

Farber, Paul Lawrence. *Finding Order in Nature: The Naturalist Tradition from Linnaeus to E. O. Wilson.* Baltimore: Johns Hopkins University Press, 2000.

Ferguson, Robert A. *Law and Letters in American Culture.* Cambridge, MA: Harvard University Press, 1984.

Fisch, Menachem, and Simon Schaffer, eds. *William Whewell: A Composite Portrait.* Oxford: Clarendon Press, 1991.

Flower, William Henry. *Essays on Museums and Other Subjects Connected with Natural History.* London: Macmillan and Co., 1898.

———. "Whale." In *Encyclopedia Britannica.* 11th ed. Cambridge: Cambridge University Press, 1910–1911.

Forster, Honore. "British Whaling Surgeons in the South Seas, 1823–1843." *Mariner's Mirror* 74, no. 4 (1988): 401–415.

———. *The South Sea Whaler.* Sharon, MA: Kendall Whaling Museum, 1985.

Foucault, Michel. *The Order of Things: An Archaeology of the Human Sciences.* New York: Pantheon Books, 1971.

Fox, Dixon Ryan. *The Decline of the Aristocracy in the Politics of New York, 1801–1840.* Edited by Robert V. Remini. New York: Harper and Row, 1965.

———. *Yankees and Yorkers.* New York: New York University Press, 1940.

Francis, John W. *Old New York, or, Reminiscences of the Past Sixty Years.* New York: C. Roe, 1858.

———. *Reminiscences of Samuel Latham Mitchill, M.D., LL.D.* New York: John F. Trow, 1859.

Frank, Stuart M. "Ballads and Songs of the American Sailor." Unpublished manuscript.

———. "Classic American Whaling Songs." Unpublished manuscript.

———. *Herman Melville's Picture Gallery: Sources and Types of the "Pictorial" Chapters of Moby-Dick.* Fairhaven, MA: Edward J. Lefkowicz, 1986.

———. *Meditations from Steerage: Two Whaling Journal Fragments.* Kendall Whaling Museum Monograph Series, no. 7. Sharon, MA: Kendall Whaling Museum, 1991.

Fraser, F. C. "An Early 17th-Century Record of the Californian Grey Whale in Icelandic Waters." *Investigations on Cetacea* 2 (1970): 13–20.

Fritz, P., and D. Williams, eds. *City and Society in the Eighteenth Century.* Toronto: University of Toronto Press, 1973.

Galison, Peter L., and D. Graham Burnett. "Einstein, Poincaré, and Modernity." *Daedalus* 132, no. 2 (2003): 41–55.

Geoffroy, Etienne, and Georges Cuvier. "Mammalogie." *Magasin Encyclopédique* 2 (1795): 152–190.

Gerbi, Antonello. *The Dispute of the New World: The History of a Polemic, 1750–1900.* Translated by Jeremy Moyle. Pittsburgh: University of Pittsburgh Press, 1973.

Gillispie, Charles Coulston. *Science and Polity in France: The Revolutionary and Napoleonic Years.* Princeton: Princeton University Press, 2004.

Golan, Tal. *Laws of Men and Laws of Nature: The History of Scientific Expert Testimony in England and America.* Cambridge, MA: Harvard University Press, 2004.

Goldsmith, Oliver. *A History of the Earth and Animated Nature.* "A New Edition." 6 vols. London: Wingrave and Collingwood, 1816.

Gotwals, Jenny. "Portfolio of a Curious Mind: The Issachar Cozzens Collection." *New-York Journal of American History* 65, no. 4 (2004): 48–57.

Gray, John Edward. *Catalogue of Seals and Whales in the British Museum.* London: Taylor and Francis, 1866.

Greene, John C. *American Science in the Age of Jefferson.* Ames: Iowa State University Press, 1984.

Guerrini, Anita. *Experimenting with Humans and Animals: From Galen to Animal Rights.* Baltimore: Johns Hopkins University Press, 2003.

Hacking, Ian. *Historical Ontology.* Cambridge, MA: Harvard University Press, 2002.

———. "Living Things." Review of *Cognitive Foundations of Natural History: Towards an Anthropology of Science,* by Scott Atran. *London Review of Books* 13 (21 February 1991): 17–18.

Hall, Courtney Robert. *A Scientist in the Early Republic: Samuel Latham Mitchill, 1764–1831.* New York: Columbia University Press, 1934.

Hall, Elton W. *Panoramic Views of Whaling by Benjamin Russell.* Old Dartmouth Historical Sketch, no. 80. New Bedford, MA: Old Dartmouth Historical Society, 1981.

Halleck, Fitz-Greene. *Fanny.* New York: C. Wiley, 1819.

Hamilton, Joseph. *Johnson's Dictionary of the English Language in Miniature*. Philadelphia: M. Carey, 1817.

Hamilton, Robert. *The Natural History of the Ordinary Cetacea or Whales*. The Naturalist's Library Series, edited by Sir William Jardine. Edinburgh: W. H. Lizars, 1837.

Haraway, Donna. *Primate Visions: Gender, Race, and Nature in the World of Modern Science*. London: Routledge, 1989.

Hardie, James. *The Description of the City of New-York*. New York: Samuel Marks, 1827.

Harland, John. *Seamanship in the Age of Sail*. Annapolis, MD: Naval Institute Press, 2000.

Harris, Jonathan. "De Witt Clinton as Naturalist." *New-York Historical Society Quarterly* 56, no. 4 (1972): 265–284.

———. "New York's First Scientific Body: The Literary and Philosophical Society, 1814–1834." *Annals of the New York Academy of Sciences* 196 (1972): 329–337.

Harris, Leslie M. *In the Shadow of Slavery: African Americans in New York City, 1626–1863*. Chicago: University of Chicago Press, 2003.

Harris, Neil. *Humbug: The Art of P. T. Barnum*. New York: Little and Brown, 1973.

Harris, Thaddeus M. *The Natural History of the Bible*. Boston: I. Thomas and E. T. Andrews, 1793.

Hartog, Hendrik. "Pigs and Positivism." *Wisconsin Law Review* 4 (1985): 899–935.

———. *Public Property and Private Power: The Corporation of the City of New York in American Law, 1730–1870*. Chapel Hill: University of North Carolina Press, 1983.

Hassler, Donald M. "Enlightenment Genres and Science Fiction: Belief and *Animated Nature* (1774)." *Extrapolation* 29, no. 4 (1998): 322–329.

Haswell, Charles. *Reminiscences of New York by an Octogenarian*. New York: Harper, 1896.

Heartman, Charles F. *The New-England Primer Issued Prior to 1830*. New York: R.R. Bowker Company, 1934.

Hedgpeth, Joel W. "*De Mirabili Maris:* Thoughts on the Flowering of Seashore Books." *Proceedings of the Royal Society of Edinburgh* (B) 72, no. 8 (1971–1972): 107–114.

Hempel, Carl Gustav. *Aspects of Scientific Explanation, and Other Essays in the Philosophy of Science*. New York: Free Press, 1965.

Heninger-Voss, Mary, ed. *Animals in Human Histories: The Mirror of Nature and Culture*. Rochester, NY: University of Rochester Press, 2002.

Hibernicus (pseud.) [DeWitt Clinton]. *Letters on the Natural History and Internal Resources of the State of New-York*. New York: E. Bliss and E. White, 1822.

Hillway, Tyrus. "Melville's Education in Science." *Texas Studies in Literature and Language* 16, no. 3 (1974): 412–425.

Hindle, Brooke. "The Underside of the Learned Society in New York, 1754–1854." In *The Pursuit of Knowledge in the Early American Republic: American Scientific and Learned Societies from Colonial Times to the Civil War*, edited by Alexandra Oleson and Sanborn C. Brown. Baltimore: Johns Hopkins University Press, 1976.

History of Fish, The. New York: Samuel Wood and Sons, 1816.

History of Wonderful Fishes and Monsters of the Ocean, A. Dublin: Graisberry and Campbell, 1816.

Hodges, Graham Russell. *Root and Branch: African Americans in New York and East Jersey, 1613–1863.* Chapel Hill: University of North Carolina Press, 1999.

Hohman, Elmo Paul. *The American Whaleman.* New York: Longmans, 1928.

Hopkins, Vivian. "The Empire State: De Witt Clinton's Laboratory." *New-York Historical Society Quarterly* 59, no. 1 (1975): 7–44.

Hosack, David. "Progress of Medical Science in New-York." *American Monthly Magazine and Critical Review* 4 (1818): 114–116.

Hough, Henry Beetle. *Wamsutta of New Bedford.* New Bedford, MA: Wamsutta Mills, circa 1946.

Howard, Mark. "Coopers and Casks in the Whaling Trade, 1800–1850." *Mariner's Mirror* 82, no. 4 (November 1996): 436–450.

Hughes, Arthur. "Science in English Encyclopedias, 1704–1875." *Annals of Science* 7, no. 4 (1951): 340–370.

Hugo, Victor. *Les Travailleurs de la Mer.* Paris: Gallimard, 1980 [1866].

Hunter, John. "Observations on the Structure and Oeconomy of Whales." *Philosophical Transactions of the Royal Society of London* 77 (1787): 371–450.

Huxley, Thomas Henry. "Owen's Position in the History of Anatomical Science." In *The Life of Richard Owen*, by Richard Owen. London: Murray, 1895.

Ingalls, Elizabeth. *Whaling Prints in the Francis B. Lothrop Collection.* Salem, MA: Peabody Museum of Salem, 1987.

Irmscher, Christoph. *The Poetics of Natural History: From John Bartram to William James.* New Brunswick, NJ: Rutgers University Press, 1999.

Irving, Washington [Diedrich Knickerbocker, pseud.]. *A History of New York.* New York: Inskeep and Bradford, 1809.

"Is a Whale a Fish?" *Port Folio* 8, no. 2, also numbered "no. 224" (1819): 129–133.

Jackson, J B S. "Dissection of a Spermaceti Whale and Three Other Cetaceans." *Boston Journal of Natural History* 5, no. 2 (1845): 137–171.

Jasanoff, Sheila. *Science at the Bar: Law, Science, and Technology in America.* Cambridge, MA: Harvard University Press, 1995.

Jefferson, Thomas. *Report of the Secretary of State on the Subject of the Cod and Whale Fisheries, Made to the House of Representatives, 1 February 1791.* Philadelphia: Francis Childe and John Swaine, 1791.

Johnson, James Weldon. *Black Manhattan.* New York: Da Capo Press, 1991 [1930].

Johnson, Samuel. *A Dictionary of the English Language.* London: W. Strahan, 1770.

Jones (pseud.) [Jackson H. Chase and George W. Heaton]. *Life and Adventure in the South Pacific. By a Roving Printer.* New York: Harper and Brothers, 1861.

Jordan, Winthrop D. *White over Black: American Attitudes toward the Negro, 1550–1812.* Chapel Hill: University of North Carolina Press, 1968.

Kelley, Wyn. *Melville's City: Literary and Urban Form in Nineteenth-Century New York.* Cambridge: Cambridge University Press, 1996.

Kettell, Samuel. *Specimens of American Poetry.* Boston: S. G. Goodrich, 1829.

Kitcher, Philip. "Species." *Philosophy of Science* 51, no. 2 (1984): 308–333.

Knickerbocker, Diedrich (pseud.) [Washington Irving]. *A History of New York.* New York: Inskeep and Bradford, 1809.

Knight, David. *Ordering the World: A History of Classifying Man.* London: Burnett Books, 1981.

Knox, Frederick John. *Account of the Rorqual, the Skeleton of Which is Now Exhibiting in the Great Rooms of the Royal Institution.* Edinburgh: A. Balfour and Co., 1835.

———. *Catalogue of Anatomical Preparations Illustrative of the Whale.* Edinburgh: Neill and Co., 1838.

Koerner, Lisbet. *Linnaeus: Nature and Nation.* Cambridge, MA: Harvard University Press, 1999.

Kohler, Robert E. *Lords of the Fly: Drosophilia Genetics and the Experimental Life.* Chicago: University of Chicago Press, 1994.

Kohlstedt, Sally Gregory. "Entrepreneurs and Intellectuals: Natural History in Early American Museums." In *Mermaids, Mummies, and Mastodons: The Emergence of the American Museum,* edited by William T. Alderson. Washington, DC: Association of American Museums, 1992.

———. "Parlors, Primers, and Public Schooling: Education for Science in Nineteenth-Century America." *Isis* 81, no. 3 (1990): 424–445.

Landauer, Lyndall Baker. "From Scoresby to Scammon: Nineteenth Century Whalers in the Foundations of Cetology." Ph.D. diss., International College (Los Angeles), 1982.

Lanman, James R. "The American Whale Fishery." *Hunt's Merchants' Magazine* 3, no. 5 (1840): 360–394.

Larson, James L. *Interpreting Nature: The Science of Living Form from Linnaeus to Kant.* Baltimore: Johns Hopkins University Press, 1994.

Lathem, Edward Connery. *Chronological Tables of American Newspapers, 1690–1820.* Worcester, MA: American Antiquarian Society, 1972.

Laurent, Goulven. *Paléontologie et Evolution en France de 1800 à 1860: Une Histoire des Idées de Cuvier et Lamarck à Darwin.* Paris: Editions du C.T.H.S., 1987.

"Leather Manufacture." *Hunt's Merchants' Magazine* 3 (1840): 141–148.

Lee, Gideon. *Two Lectures on Tanning.* New York: The Eclectic Fraternity, 1838.

Lewis, Andrew John. "The Curious and the Learned: Natural History in the Early Republic." Ph.D. diss., Yale University, 2001.

Lightman, Bernard, ed. *The Dictionary of Nineteenth-Century British Scientists.* 4 vols. Chicago: University of Chicago Press, 2004.

———, ed. *Victorian Science in Context.* Chicago: University of Chicago Press, 1997.

Linnaeus, Carl. *Systema Naturae.* 9th ed. Leiden: Haak, 1756.

———. *Systema Naturae.* 10th ed. 2 vols. Stockholm: Salvius, 1758.

———. *Systema Naturae.* 12th ed. 3 vols. Stockholm: Salvius, 1766.

Lipton, Peter. *Inference to the Best Explanation.* London: Routledge, 1991.

Looby, Christopher. "The Constitution of Nature: Taxonomy as Politics in Jefferson, Peale, and Bartram." *Early American Literature* 22, no. 3 (1987): 252–273.

Losee, John. "Whewell and Mill on the Relation between Philosophy of Science and History of Science." *Studies in the History and Philosophy of Science* 14, no. 2 (1983): 113–126.

Lovejoy, Arthur O. *The Great Chain of Being: The Study of the History of an Idea.* Cambridge, MA: Harvard University Press, 1966 [1936].

Lund, Roger D. "The Eel of Science: Index Learning, Scriblerian Satire, and the Rise of Information Culture." *Eighteenth-Century Life* 22, no. 2 (1998): 18–42.

Lyell, Charles. *Principles of Geology.* 3 vols. London: John Murray, 1830-1833.

Mackenzie, John M. *The Empire of Nature: Hunting, Conservation, and British Imperialism.* Manchester: Manchester University Press, 1988.

Maehle, Andreas-Holger. "Cruelty and Kindness to the 'Brute Creation': Stability and Change in the Ethics of the Man-Animal Relationship, 1600–1850." In *Animals and Human Society,* edited by Aubrey Manning and James Serpell. London: Routledge, 1994.

Malloy, Mary. "Whalemen's Perceptions of 'The High and Mighty Business of Whaling.'" *Log of Mystic Seaport* 41, no. 2 (1989): 56–67.

Manning, Aubrey, and James Serpell, eds. *Animals and Human Society: Changing Perspectives.* London: Routledge, 1994.

Martin, Kenneth R. *Whalemen's Paintings and Drawings: Selections from the Kendall Whaling Museum Collection.* Sharon, MA: Kendall Whaling Museum, 1983.

Maury, Matthew Fontaine. *Explanations and Sailing Directions to Accompany the Wind and Current Charts.* 7th ed. Philadelphia: E. C. and J. Biddle, 1855.

Mavor, William. *Catechism of Animated Nature, or, An Easy Introduction to the Animal Kingdom: For the Use of Schools and Families.* New York: Samuel Wood and Sons, 1819.

McConnell, Anita. "The Scientific Life of William Scoresby Jnr., with a Catalogue of His Instruments and Apparatus in the Whitby Museum." *Annals of Science* 43 (1986): 257–286.

McKay, Richard C. *South Street: A Maritime History of New York.* New York: Putnam, 1934.

McLachlan, James. "The Choice of Hercules: American Student Societies in the Early 19th Century." In *The University in Society,* vol. 2, edited by Lawrence Stone. Princeton: Princeton University Press, 1974.

McMartin, Barbara. *Hides, Hemlocks and Adirondack History: How the Tanning Industry Influenced the Region's Growth.* Utica, NY: North Country Books, 1992.

McOuat, Gordon. "Cataloguing Power: Delineating 'Competent Naturalists' and the Meaning of Species in the British Museum." *British Journal for the History of Science* 34 (2001): 1–28.

———. "From Cutting Nature at Its Joints to Measuring It: New Kinds and New Kinds of People in Biology." *Studies in the History and Philosophy of Science* 32, no. 4 (2001): 613–645.

Melville, Herman. *Melville* [Selections]. New York: Library of America, 1984.

———. *Moby-Dick, or, The Whale*. Edited by Harrison Hayford, Hershel Parker, and G. Thomas Tanselle. Evanston, IL: Northwestern University Press and the Newberry Library, 1988 [1851].

Melville, Richard V. *Towards Stability in the Names of Animals: A History of the International Commission on Zoological Nomenclature, 1895–1995*. London: International Trust for Zoological Nomenclature, 1995.

Merton, Robert K. *On the Shoulders of Giants: A Shandean Postscript*. New York: Free Press, 1965.

Mill, John Stuart. *A System of Logic, Ratiocinative and Inductive*. New York: Harper and Brothers, 1846.

Millender, Michael Jonathan. "The Transformation of the American Criminal Trial, 1790–1875." Ph.D. diss., Princeton University, 1996.

Miller, Douglas T. *Jacksonian Aristocracy: Class and Democracy in New York, 1830–1860*. New York: Oxford University Press, 1967.

Miller, Lillian B., and David C. Ward, eds. *New Perspectives on Charles Willson Peale*. Pittsburgh: University of Pittsburgh Press and the Smithsonian Institution, 1991.

Miller, Perry. *Nature's Nation*. Cambridge, MA: Harvard University Press, 1967.

———. *The Raven and the Whale: Poe, Melville, and the New York Literary Scene*. Baltimore: Johns Hopkins University Press, 1997 [1956].

Mitchill, Samuel Latham. "Additional Proof of the Existence of Large Animals in the Ocean." *The Medical Repository* 17 (1813–1814): 388–390.

———. "Cod Fishery in the United States." *The Medical Repository* 8 (1804–1805): 87.

———. "Detailed Abstract of the French Professor Duméril's System of Zoölogy." *The Medical Repository* 10 (1806–1807): 156–164.

———. "Effects of the Greenland Ice on the Atlantic Ocean." *The Medical Repository* 16 (1812–1813): 194–196.

———. "An Exhibition of Facts, Showing the Progress of Ice Islands." *The Medical Repository* 10 (1806–1807): 225–235.

———. "Exhibition of Several Wrong Ideas, Whereby Medical and Chemical Knowledge Have Been Remarkably Perverted." *The Medical Repository* 5 (1801–1802): 113–123.

———. "Experiments, Facts and Observations in Natural History." *The Medical Repository* 5 (1801–1802): 205–240.

———. "Experiments on the Production of Watery Vapour." *The Medical Repository* 4 (1800–1801): 309–312.

———. "Extension of the Means of Relief to Fredish Seamen." *The Medical Repository* 8 (1804–1805): 423–433.

———. "Facts and Observations Relative to the Trade from New-York to the Fiji Islands." *The Medical Repository* 14 (1810–1811): 209–215.

———. "Facts and Observations Showing the Existence of Large Animals in the Ocean." *The Medical Repository* 16 (1812–1813): 396–407.

———. "Facts Concerning the Generation of Eels." *The Medical Repository* 10 (1806–1807): 201–203.

———. "Fishes of New-York." *The Medical Repository* 17 (1813–1814): 280–294.

———. "The Fishes of New-York." *Transactions of the Literary and Philosophical Society of New-York* 1 (1815): 355–492.

———. "Is a Whale a Fish?" *The New-York Literary Journal, and Belles-Lettres Repository* 3, no. 1 (1820): 60–61.

———. "Letter to C.W. Peale Regarding Alexander Wilson." *The Medical Repository* 17 (1813–1814): 251–252.

———. "Memoir on Ichthyology: The Fishes of New-York, described and arranged. In a supplement to the Memoir on the same subject." *American Monthly Magazine and Critical Review* 2 (1818): 241–248.

———. "Memoir on Ichthyology (Continued)." *American Monthly Magazine and Critical Review* 2 (1818): 321–328.

———. "Mode of Destroying Printer's Ink by Counterfeiters." *The Medical Repository* 4 (1800–1801): 322–324.

———. "Outline of Professor Mitchill's Lectures in Natural History in the College at New-York, delivered in 1809–1810, previous to his departure for Albany, to take his seat in the Legislature of the State." *The Medical Repository* 13 (1809–1810): 257–267.

———. "An Ovo-viviparous Animal ... the Anatomy and Physiology of the Shark." *The Medical Repository* 8 (1804–1805): 78–81.

———. *Picture of New-York, or, The Traveller's Guide through the Commercial Metropolis of the United States.* New York: I. Riley, 1807.

———. *Some of the Memorable Events and Occurrences in the Life of Samuel L. Mitchill of New-York, from the Year 1786 to 1826.* New York: n.p., 1828.

———. "Views of the Manufactures in the United States." *The American Medical and Philosophical Register* 2 (April 1812): 405–413.

Mizelle, Brett. "'Man Cannot Behold It Without Contemplating Himself': Monkeys, Apes and Human Identity in the Early Republic." *Pennsylvania History: A Journal of Mid-Atlantic Studies* 66 (1999): 145–173.

Mohr, James C. *Doctors and the Law: Medical Jurisprudence in Nineteenth-Century America.* Baltimore: Johns Hopkins University Press, 1993.

Moment, David. "The Business of Whaling in America in the 1850's." *The Business History Review* 31, no. 3 (1957): 261–291.

Morfit, Campbell. *The Arts of Tanning, Currying and Leather-Dressing.* Philadelphia: Henry Carey Baird, 1852.

Morris, Richard B. *Select Cases of the Mayor's Court of New York City 1674–1784.* Washington, DC: American Historical Association, 1935.

Mott, Frank Luther. *A History of American Magazines.* 5 vols. Cambridge, MA: Harvard University Press, 1938.

Moyer, Albert E. *Joseph Henry: The Rise of an American Scientist.* Washington, DC: Smithsonian Institution, 1997.

Neill, Patrick. *Some Account of a Fin-Whale, Stranded near Alloa.* Edinburgh [?]: n.p.,
    circa 1810.

"New-York Institution." *American Monthly Magazine and Critical Review* 1 (1817):
    271–273.

New York (State) Legislature. *Journal of the Assembly of the State of New-York: At Their
    Forty-Second Session.* Albany: J. Buel, 1819.

Nicholson, William. *The British Encyclopedia, or Dictionary of Arts and Sciences.* 6 vols.
    London: C. Whittingham, 1809.

Nodyne, Kenneth R. "The Founding of the Lyceum of Natural History." *Annals of
    the New York Academy of Sciences* 172 (1970): 141–149.

———. "The Role of De Witt Clinton and the Municipal Government in the De-
    velopment of Cultural Organizations in New York City, 1803–1817." Ph.D. diss.,
    New York University, 1969.

Norcross, Frank W. *A History of the New York Swamp.* New York: Chiswick Press,
    1901.

Oleson, Alexandra, and Sanborn C. Brown, eds. *The Pursuit of Knowledge in the Early
    American Republic: American Scientific and Learned Societies from Colonial Times
    to the Civil War.* Baltimore: Johns Hopkins University Press, 1976.

Olmsted, Francis Allyn. *Incidents of a Whaling Voyage.* New York: D. Appleton and
    Co., 1841.

"On the Means of Education and the Scientific Institutions in New York." *Analectic
    Magazine* 13 (1819): 452–459.

Orosz, Joel J. *Curators and Culture: The Museum Movement in America, 1740–1870.*
    Tuscaloosa: University of Alabama Press, 1990.

Otter, Samuel. *Melville's Anatomies.* Berkeley: University of California Press, 1999.

Outram, Dorinda. *Georges Cuvier: Vocation, Science, and Authority in Post-Revolutionary
    France.* Manchester: Manchester University Press, 1984.

Owen, Richard. *The Life of Richard Owen.* 2 vols. London: Murray, 1895.

Parker, Hershel. *Herman Melville: A Biography.* 2 vols. Baltimore: Johns Hopkins
    University Press, 1996–2002.

Parley, Peter [pseud.]. *Parley's Penny Library, or, Treasury of Knowledge, Entertainment
    and Delight.* London: Cleave, 1841.

Parrish, Susan Scott. *American Curiosity: Cultures of Natural History in the Colonial
    British Atlantic World.* Chapel Hill: University of North Carolina Press, 2006.

Pascalis, Felix. *Eulogy on the Life and Character of the Hon. Samuel Latham Mitchill,
    M.D.* New York: American Argus Press, 1831.

Pauly, Philip. *Biologists and the Promise of American Life.* Princeton: Princeton Uni-
    versity Press, 2002.

Peale, Charles Willson. *The Collected Papers of Charles Willson Peale and His Family,
    1735–1885.* 449 microfiche cards. Millwood, NY: KTO Microform, 1980.

Peale, Rembrandt. "A Short Account of the Mammoth." *Philosophical Magazine* 14
    (1803): 162–169.

Pennant, Thomas. *History of Quadrupeds.* 2 vols. London: B. White, 1781.

———. *History of Quadrupeds.* 3rd ed. 2 vols. London: B. and J. White, 1793.

Philbrick, Nathaniel. *In the Heart of the Sea: The Tragedy of the Whaleship* Essex. New York: Viking, 2000.

Phillipson, Nicholas T. "Towards a Definition of the Scottish Enlightenment." In *City and Society in the Eighteenth Century,* edited by P. Fritz and D. Williams. Toronto: University of Toronto Press, 1973.

Pilleri, Georg, and Lucie Arvy. *The Precursors in Cetology: From Guillaume Rondelet to John Anderson.* Berne: Brain Anatomy Institute, 1981.

Pintard, John. *Letters from John Pintard.* 4 vols. New York: New-York Historical Society, 1940.

Pitman, James Hall. *Goldsmith's Animated Nature: A Study of Goldsmith.* New Haven: Yale University Press, 1924.

Porter, Charlotte M. *The Eagle's Nest: Natural History and American Ideas, 1812–1842.* Tuscaloosa: University of Alabama Press, 1986.

Post, Francis. "Natural History of the Spermaceti Whale." In *Explanations and Sailing Directions to Accompany the Wind and Current Charts,* by Matthew Fontaine Maury, 7th ed. Philadelphia: E. C. and J. Biddle, 1855.

Prince, Sue Ann, ed. *Stuffing Birds, Pressing Plants, Shaping Knowledge: Natural History in North America, 1730–1860.* Philadelphia: American Philosophical Society, 2003.

Putnam, Hilary. *Mind, Language, and Reality.* Cambridge: Cambridge University Press, 1975.

Rader, Karen. *Making Mice: Standardizing Animals for American Biomedical Research, 1900–1955.* Princeton: Princeton University Press, 2004.

Rafinesque, Constantine S. "Introduction to the Ichthyology of the United States." *American Monthly Magazine and Critical Review* 2 (1818): 202–207.

Ray, John. *Synopsis Methodica Animalium Quadrupedum et Serpentini Generis.* London: Southwell, 1693.

Rediker, Marcus. *Between the Devil and the Deep Blue Sea.* Cambridge: Cambridge University Press, 1987.

Rees, Abraham. *The Cyclopaedia or Universal Dictionary of Arts, Sciences, and Literature.* 47 vols. Philadelphia: Samuel F. Bradford and Murray, Fairman and Co., 1805–1825.

Regis, Pamela. *Describing Early America: Bartram, Jefferson, Crèvecoeur, and the Rhetoric of Natural History.* DeKalb: Northern Illinois University Press, 1992.

Reynolds, Jeremiah N. *Address, on the Subject of a Surveying and Exploring Expedition.* New York: Harper and Brothers, 1836.

———. "Mocha Dick or the White Whale of the Pacific." *Knickerbocker Magazine* 13 (May 1839): 377–392.

Rigal, Laura. *The American Manufactory: Art, Labor, and the World of Things in the Early Republic.* Princeton: Princeton University Press, 1998.

Ritvo, Harriet. *The Animal Estate: The English and Other Creatures in Victorian England.* Cambridge, MA: Harvard University Press, 1987.

———. *The Platypus and the Mermaid, and Other Figments of the Classifying Imagination.* Cambridge, MA: Harvard University Press, 1997.

Robertson, James. "Description of the Blunt-Headed Cachalot." *Philosophical Transactions of the Royal Society* 60 (1770): 321–324.

Rogers, Daniel. *The New-York City-Hall Recorder, for the Year 1818.* New York: Abraham Vosburgh, 1818.

Rolfe, W. D. Ian. "William and John Hunter: Breaking the Great Chain of Being." In *William Hunter and the Eighteenth-Century Medical World,* edited by W. F. Bynum and Roy Porter. Cambridge: Cambridge University Press, 1985.

Roman, Joe. *Whale.* London: Reaktion, 2006.

Rosner, Lisa. *Medical Education in the Age of Improvement: Edinburgh Students and Apprentices, 1760–1826.* Edinburgh: Edinburgh University Press, 1991.

———. "Thistle on the Delaware: Edinburgh Medical Education and Philadelphia Practice, 1800–1825." *Social History of Medicine* 5, no. 1 (1992): 19–42.

Rothfels, Nigel. *Savages and Beasts: The Birth of the Modern Zoo.* Baltimore: Johns Hopkins University Press, 2002.

Rupke, Nicolaas A., ed. *Vivisection in Historical Perspective.* London: Croon Helm, 1987.

Rutland, Robert A., ed. *James Madison and the American Nation, 1751–1836: An Encyclopedia.* New York: Simon and Schuster, 1994.

Sampson, William. *Is a Whale a Fish? An Accurate Report of the Case of James Maurice against Samuel Judd.* New York: Van Winkle, 1819.

Scammon, Charles M. *The Marine Mammals of the Northwestern Coast of North America.* San Francisco: John H. Carmany and Co., 1874.

Schaffer, Simon. "The History and Geography of the Intellectual World: Whewell's Politics of Language." In *William Whewell: A Composite Portrait,* edited by Menachem Fisch and Simon Schaffer. Oxford: Clarendon Press, 1991.

———. "A Science Whose Business Is Bursting: Soap Bubbles as Commodities in Classical Physics." In *Things That Talk: Object Lessons from Art and Science,* edited by Lorraine Daston. Cambridge, MA: Zone, 2004.

———. "The Urbanity of Classical Physics: Soap Bubbles, Time Machines and Other Metropolitan Marvels." Lawrence Stone Visiting Professorship Lecture, Princeton University, 9 December 2003. Unpublished paper.

Schiebinger, Londa. *Nature's Body: Gender in the Making of Modern Science.* Boston: Beacon, 1993.

Scholnick, Robert J., ed. *American Literature and Science.* Lexington: University Press of Kentucky, 1992.

Scoresby, William. *An Account of the Arctic Regions with a History and Description of the Northern Whale Fishery.* 2 vols. London: Constable, 1820.

———. "Remarks on the Size of the Greenland Whale." *The Edinburgh Philosophical Journal* 1 (1819): 83–88.

Secord, James. *Victorian Sensation: The Extraordinary Publication, Reception, and Secret Authorship of* Vestiges of the Natural History of Creation. Chicago: University of Chicago Press, 2000.

Sedgwick, Theodore. *A Treatise on the Rules Which Govern the Interpretation and Application of Statutory and Constitutional Law.* New York: J. S. Voorhies, 1857.

Semonin, Paul. *American Monster: How the Nation's First Prehistoric Creature Became a Symbol of National Identity.* New York: New York University Press, 2000.

———. "'Nature's Nation': Natural History as Nationalism in the New Republic." *Northwest Review* 30 (1992): 6–41.

Shaw, George. *General Zoology, or Systematic Natural History.* 14 vols. London: G. Kearsley, 1800–1826.

———. *The Naturalist's Miscellany.* 24 vols. London: Nodder and Co., 1789–1813.

Sheridan, Richard Brinsley. *The Critic.* Edited by David Crane. London: A. and C. Black, 1989.

Sherman, Stuart C. *Whaling Logbooks and Journals, 1613–1927: An Inventory of Manuscript Records in Public Collections.* Rev. ed. New York: Garland, 1986.

Shoemaker, Nancy. "Whale Meat in American History." *Environmental History* 10, no. 2 (2005): 269–294.

"Sketch of the Life and Character of the Late Gideon Lee." *Hunt's Merchants' Magazine* 8 (1843): 57–64.

Smallwood, William Martin. *Natural History and the American Mind.* New York: Columbia University Press, 1941.

Smith, Edgar Fahs. *Samuel Latham Mitchill: A Father in American Chemistry.* New York: Columbia University Press, 1922.

Smith, Richard Dean. *Melville's Science: "Devilish Tantalization of the Gods!"* New York: Garland, 1993.

Smith, Roger, and Brian Wynne, eds. *Expert Evidence: Interpreting Science in the Law.* London: Routledge, 1989.

Smyth, Albert Henry. *Philadelphia Magazines and Their Contributors, 1741–1850.* Philadelphia: R. M. Lindsay, 1892.

Snyder, Laura J. "The Mill-Whewell Debate: Much Ado about Induction." *Perspectives on Science* 5, no. 2 (1997): 159–198.

Sorber, Elliott. "Sets, Species, and Evolution: Comments on Philip Kitcher's 'Species.'" *Philosophy of Science* 51, no. 2 (1984): 334–341.

Stackpole, Eduard A. *Whales and Destiny: The Rivalry between America, France, and Britain for Control of the Southern Whale Fishery, 1785–1825.* Amherst: University of Massachusetts Press, 1972.

Stamp, Tom, and Cordelia Stamp. *William Scoresby, Arctic Scientist.* Whitby, UK: Caedmon, 1976.

Starbuck, Alexander. *History of the American Whale Fishery from Its Earliest Inception to the Year 1876.* Waltham, MA: Published by the Author, 1878.

Stevenson, Louise L. "Preparing for Public Life: the Collegiate Students at New York University, 1832–1881." In *The University and the City,* edited by Thomas Bender. New York: Oxford University Press, 1988.

Stokes, Isaac Newton Phelps. *The Iconography of Manhattan Island, 1498–1909.* 6 vols. New York: Robert H. Dodd, 1915–1928.

Stone, Lawrence, ed. *The University in Society*. 2 vols. Princeton: Princeton University Press, 1974.

Strum, Harvey. "Property Qualifications and Voting Behavior in New York, 1807–1816." *Journal of the Early Republic* 1 (1981): 347–371.

Thackray, Arnold. "Natural Knowledge in Cultural Context: The Manchester Model." *American Historical Review* 79, no. 3 (1974): 672–709.

Thomas, Keith. *Man and the Natural World: Changing Attitudes in England, 1500–1800*. Oxford: Oxford University Press, 1983.

Todes, Daniel P. "Pavlov's Physiology Factory." *Isis* 88, no. 2 (1997): 205–246.

Tønnessen, J. N., and Arne Odd Johnsen. *The History of Modern Whaling*. Translated by R. I. Christophersen. Berkeley: University of California Press, 1982.

Tower, Walter S. *A History of the American Whale Fishery*. Philadelphia: University of Pennsylvania Press, 1907.

Townsend, Peter S. *An Anniversary Discourse, Delivered before the Lyceum of Natural History of New-York*. New York: C. Wiley and Co., 1820.

Turner, H. N. "An Essay on Classification." *Zoologist* 5 (1847): 1943–1955.

Turner, James C. *Reckoning with the Beast: Animals, Pain, and Humanity in the Victorian Mind*. Baltimore: Johns Hopkins University Press, 1980.

U.S. Congress. House. *Report of Honorable Thomas Jefferson, Secretary of State on the Subject of the Cod and Whale Fisheries, Made to the House of Representatives*. 42nd Cong., 2nd sess., Mis. Doc. No. 32. Printed 8 January 1872 [originally presented 1 February 1791].

Verplanck, Gulian C. *The State Triumvirate, A Political Tale*. New York: For the Author, 1819.

Vincent, Howard P. *The Trying-out of Moby-Dick*. Boston: Houghton Mifflin Co., 1949.

Walsh, Walter J. "Redefining Radicalism: A Historical Perspective." *George Washington Law Review* 59 (1991): 636–682.

Warren, Leonard. *Constantine Samuel Rafinesque: A Voice in the American Wilderness*. Lexington: University Press of Kentucky, 2004.

Weiss, Harry B. *Whaling in New Jersey*. Trenton: New Jersey Agricultural Society, 1974.

Welsh, Peter C. "A Craft That Resisted Change: American Tanning Practices to 1850." *Technology and Culture* 4 (1963): 299–317.

———. *Tanning in the United States to 1850*. Washington: Smithsonian Institution, 1964.

Wheeler, Jacob D. *Reports of Criminal Law Cases*. With notes and references by Thomas W. Waterman. New York: Banks and Brothers, 1860.

Whewell, William. *The Philosophy of the Inductive Sciences, Founded upon Their History*. vol. 1, *The Language of Science*. London: John W. Parker, 1840.

White, Paul. "The Experimental Animal in Victorian Britain." In *Thinking with Animals: New Perspectives on Anthropomorphism*, edited by Lorraine Daston and Gregg Mitman. New York: Columbia University Press, 2004.

White, Richard. *The Organic Machine.* New York: Hill and Wang, 1995.

White, Shane. *Somewhat More Independent: The End of Slavery in New York City, 1770–1810.* Athens: University of Georgia Press, 1991.

Whitehead, William A., Frederick William Ricord, William Nelson, et al., eds. *Documents Relating to the Colonial History of the State of New Jersey.* 35 vols. Patterson, Bayonne, etc., NJ: various presses, 1880–1899.

Widmer, Edward L. *Young America: The Flowering of Democracy in New York City.* New York: Oxford University Press, 1999.

Wilentz, Sean. *Chants Democratic: New York City and the Rise of the American Working Class, 1788–1850.* New York: Oxford University Press, 1984.

Willis, Roy G., ed. *Signifying Animals: Human Meaning in the Natural World.* London: Unwin Hyman, 1990.

Willughby, Francis, and John Ray. *De Historia Piscium.* Oxford: Sheldon, 1686.

Wilson, David Scofield. *In the Presence of Nature.* Amherst: University of Massachusetts Press, 1978.

Wilson, John. *The Cruise of the "Gipsy": The Journal of John Wilson, Surgeon on a Whaling Voyage to the Pacific Ocean, 1839–1843.* Edited by Honore Forster. Fairfield, WA: Ye Galleon Press, 1991.

Winsor, Mary P. "Barnacle Larvae in the Nineteenth Century: A Case Study in Taxonomic Theory." *Journal of the History of Medicine* 24 (1969): 295–309.

———. *Reading the Shape of Nature: Comparative Zoology at the Agassiz Museum.* Chicago: University of Chicago Press, 1991.

———. *Starfish, Jellyfish, and the Order of Life: Issues in Nineteenth-Century Science.* New Haven: Yale University Press, 1976.

Winter, Alison. "Compasses All Awry: The Iron Ship and the Ambiguities of Cultural Authority in Victorian Britain." *Victorian Studies* 38 (1994): 69–98.

Winterer, Caroline. *The Culture of Classicism: Ancient Greece and Rome in American Intellectual Life, 1780–1910.* Baltimore: Johns Hopkins University Press, 2002.

Woodworth, Samuel. *Beasts at Law, or, Zoologian Jurisprudence.* New York: J. Harmer and Co., 1811.

Wright, Nathalia. "Moby Dick: Jonah's or Job's Whale?" *American Literature* 37, no. 2 (1965): 190–195.

Yeo, Richard R. *Defining Science: William Whewell, Natural Knowledge, and Public Debate in Early Victorian Britain.* Cambridge: Cambridge University Press, 1993.

———. *Encyclopaedic Visions: Scientific Dictionaries and Enlightenment Culture.* Cambridge: Cambridge University Press, 2001.

———. "Reading Encyclopedias: Science and the Organization of Knowledge in British Dictionaries of Arts and Sciences, 1730–1850." *Isis* 82, no. 1 (1991): 24–49.